What I Didn't Learn in Flying School

This book is a straightforward guide
to the professional upgrading of pilot skills;
preflight to parking, student through ATP.
The author has fifty-plus years experience
flying and instructing in general aviation,
military and major airline operations.

by
Vernon M. (Vern) Foster

Library of Congress Catalog Card Number 95-92050
ISBN 0-9644332-0-6
Printed in the United States of America
First Printing 1995

Corona Pass Aviation, Inc.
P.O. Box 440799 Aurora, Colorado 80044-0799

To my wife Kay,
who has endured 47 years
as a pilot's wife.

Caution to the Reader

The factual description of each accident recounted in this book is drawn from the National Transportation Safety Board's report of the accident.

These accidents are selected to illustrate particular safety considerations being discussed and are not intended to judge or reach any definitive conclusions about the ability or capacity of any person involved, whether living or dead, or the quality of any aircraft or accessory involved.

For specific guidance and advice in developing personal flight safety habits, the author encourages the reader to consult your own flight instructor and the P.O.H. for the aircraft being operated.

WHAT I DIDN'T LEARN IN FLYING SCHOOL

TABLE OF CONTENTS

TABLE OF CONTENTS

ACKNOWLEDGEMENTS

My special thanks to our granddaughter Renee and husband Paul Harvel for nearly two years of word processing, page set-up, advice and many revisions. Their work and patience are greatly appreciated.

Thanks to Dean Moberg who worked patiently thru several revisions to develop the figures in this book.

Carl M. Williams, my pilot/businessman friend of many years, who provided extraordinary support and encouragement to *"get some of that good stuff written down"!*

Thanks to my fellow CFI Bill Marshall for the cover photos.

Thanks to our daughter Linda and husband Gary, and our son, CDR. W.K. (Bill) Foster USN Retired and wife Judy for their encouragement.

A tip of the hat to my many men and women pilot and instructor friends, who cheered me on.

Special recognition to Colorado Pilots Association for written permission to use materials from the ground school manual entitled **MOUNTAIN FLYING**.

Thanks to Tony Pierce for the Cover Design.

ABOUT THE AUTHOR

Vern Foster had his first flying lesson at Wichita, Kansas, in 1939. He was a 1942 graduate of the U.S. Army Air Corps Central Instructors School at Kelly Field, Texas. His first flying job was an Air Corps Primary Flight Instructor instructing in open cockpit PT-19 aircraft.

Then followed a career of 52 years which included Trans-Pacific Air Transport Command pilot, charter pilot and 34 years with United Airlines as pilot and instructor until retirement in 1984. In parallel with these jobs, he has actively pursued general aviation flight instruction from 1945 to the present -- mid 1995. He has instructed in more than 87 makes and models of general aviation aircraft. These include virtually all singles and light twins manufactured since 1946. Others are Stearman, Ryan PT-22, Staggerwing Beech, Twin Beech, Cessna 195, Gullwing Stinsons and other pre-WWII aircraft.

In the early 1960's he developed FAA/VA approved ground and flight courses through ATP. He taught aviation classes at Metro State College in Denver during the 1970's. More recently he was a major developer of the text entitled **MOUNTAIN FLYING** which is the basis of the Colorado Pilots Association widely acclaimed Mountain flying course.

In 1987 he was inducted into the *COLORADO AVIATION HALL of FAME*. He is a member of the *BLUE MAX* society. He is a past President of Colorado Pilots Association and a long time FAA Accident Prevention Counselor. He has CFII, SEL, MEL, SES and ATP which was originally issued in 1958.

PREFACE

Have you ever come across a new technique, procedure, idea or refinement which made a sport, a hobby or your job easier, more accurate or safer? You probably have and your likely response has been like mine -- **WHY DIDN'T SOMEONE TELL ME THIS BEFORE NOW?**

My love of flying and my pride in piloting and instructing was ongoing motivation to build the basic skills of my early training. Thus, thanks to a career of general aviation, military and airline flying and instructing I have developed this "HOW TO..." book of extra training which I believe every pilot should master. As you study this book I will be truly gratified if you exclaim again and again -- *"WHY DIDN'T SOMEONE TELL ME THIS BEFORE NOW?"*

This book addresses the 15 operational areas which I believe to be in the greatest need of upgrading and refinement. These include mountain flying, night flying, IFR operations, tailwheel flying and precision VFR approaches to mention a few. There is a complete chapter covering each of these 15 important and rarely refined areas of flying. The chapters move from pre-flight thru the landing so that continuity and practicality enhance the understanding and retention of the material.

A professional attitude and knowledge coupled with a disciplined plan of operation are the foundation of being a good pilot no matter how many or how few flying hours one may have. Look at this as The Art of Flying!

Each of us must continuously strive to stay abreast of the ongoing progress of the piloting job. There is a common misconception that, once a pilot has acquired an FAA pilot certificate, he or she has *"learned to fly"*. Wrong!

PREFACE

Pilot certificates should be looked upon as *"licenses to learn"*, from ongoing flying experiences, NTSB accident reports, books, articles in flying magazines, worthwhile input from CFI's and the experiences of fellow pilots. Keep in mind that flying techniques and procedures are in a process of continuous change and improvement. But, general aviation is lagging far behind the military and the airlines in overall professionalism.

Imagine an airline crew trying to operate an MD-80, B-737 or B-747 using techniques and procedures which are unchanged for more than 50 years! Sounds a little scary doesn't it? However, the basic requirements for a Private License and thus the associated training have changed very little since 1939. Over the years attempts to expand the training requirements have been met by vigorous protests from industry and pilot groups. My main suggestions for improvement are to at least double and revise the cross country training requirement. Additionally, a fresh method of teaching and flying VFR approaches should be introduced. These and many other items are covered in this book. The bottom line is that general aviation pilot training needs to be upgraded to adequately prepare the general aviation pilot to operate the current mix of aircraft in today's operating environment.

It is my opinion and the opinion of many others that the current objective is to get a pilot certificate rather than learn to operate safely and efficiently in the real world outside the local area.

This book contains several condensations of National Transportation Safety Board accident reports. These are presented only with the intent that pilots may learn from the experiences of other pilots. My perceptions and critique follow some of these reports.

Many pilots appear to be *"stuck"* at the level of knowledge and piloting ability which existed at the time the pilot certificate was issued. I know from giving BFR'S that there are far too many people in this category. They seem to be saying -- *"hey, I have a pilot's license, I do O.K."*, etc. These are the folks most prone to accidents.

After my initial flight instructor training I assumed that I was ready to be a flight instructor. This training consisted of a Civilian Pilot Training Secondary Flight Instructor course at Lawrence, Kansas. This involved primary maneuvers plus 35 hours in Waco UPF-7's in precision maneuvers and aerobatics and topped off with 5 hours of night flying in a Fairchild 24. Later I graduated from the U.S. Army Air Corps Central Flight Instructors School at Kelly Field, Texas. This program was all flown in the PT-19 and covered the full range of the primary program from zero time through aerobatics. I was then assigned to an Air Corps Primary training base in Texas. During the next few months I worked hard to adapt to the real world of doing the flight instructor job. I learned from trainees, fellow instructors and the various experiences involved in day to day operations.

With 52 years of flight instructor experience of which 50 years are in 87 makes/models of general aviation aircraft, I have had a lot of great experiences and I've learned alot. The purpose of this book is to pass along some of this information to my fellow pilots and CFI's with the hope that they will benefit.

I encourage each reader to question any techniques and procedures if you have any doubt of their logic or validity. For example, I find that very few pilots and instructors really understand the different types of wing flaps and the advantages of each.

Pilots are frequently operating contrary to the POH in a variety of situations. A pilot may fly one make or model of aircraft using the correct procedures then switch to a different make/model and use the same practices which are improper or possibly unsafe in this aircraft. In many instances doing it wrong will work out. But don't count on it.

In the world of light twins the lack of understanding or outright wrong information is widespread. Most pilots have never heard of *"drift down"* as applied to multi-engine operations much less understand it. Pitch attitude flying has been largely ignored. In the operation of tailwheel aircraft the misinformation is also widespread. Lastly, knowledge about mountain flying is virtually nonexistent except in a few areas of the western U.S.. This book contains a comprehensive chapter on mountain flying.

Some of what I advocate in this book may be at odds with what you believe, have heard, or been taught. There is lots of wrong information and operating practices out there which are leading to unsafe operations and often to accidents. The objective of this book is to upgrade and enhance your piloting operations.

Many of the items in this book are developments which I have made over the years to make flying and teaching easier and more precise. In other items I have expanded upon or revised good long standing practices. In several chapters the text is supported by appropriate figures and graphs.

Everything in this book will stand the test of operational logic. It took me many years of instructing and flying to finally figure out how to *"do it right"* the first time -- most of the time.

Thus the title of this book -- **WHAT I DIDN'T LEARN IN FLYING SCHOOL.** That is a bit of a stretch, of course, because I did learn alot in flying school -- but I largely taught myself to be an instructor and better pilot. I invite you to do the same.

Regardless of your pilot certificates, ratings, hours or years of flying experience, I trust that you will find something of benefit within the covers of the book. So get into a comfortable spot and enjoy what I trust will be the most productive flight instructor briefing to come your way!

"Good flying and keep the greasy side down!"

Vern Foster.

PILOT IMPAIRMENT

Are you physically and mentally prepared to fly? Pilot impairment contributes to more accidents than failures of engines, aircraft and aircraft systems. You must be honest when making this judgement and never fly when impaired. An impairment is any mental or physical condition that distracts from your ability to combine rapid and accurate assessments with the appropriate responses.

FATIGUE

Flying after a long period of physical or mental exertion sets the stage for actions or inactions which lead to accidents. Records show many instances of approach and landing accidents occurring late in the day after exhaustive business or pleasure activities. Avoid long nonstop flights. I consider a flight leg of 3.0 to 3.5 hours to be the maximum length for safe pilot operations. We cannot stand or move around in single engine and light twin aircraft. Our bodies need the exercise of standing and walking, not to mention the needed pit stop. Our minds need a short period of relaxation. Being aloft when you are thirsty, hungry or fatigued is unpleasant for pilot and passengers but, more importantly, it is hazardous. Let's leave those long non-stops to the airline and military transport crews.

STRESS

Flying when heavily impacted by things such as a divorce, death in the family, a heated argument at home or at work, severe financial or on the job problems can be hazardous. If you are involved with these kinds of situations, delay the flight or avail yourself of alternate means of transportation. Don't start the engine unless you can put your full attention to flying.

ALCOHOL

One of my dear friends who retired as a B-747 Captain had a rule on drinking which I believe is a great guideline; NO alcoholic beverage of any kind during the 24 hour period prior to flight! Sounds pretty simple, doesn't it? He enjoyed a drink at appropriate times but his sense of professionalism would not allow for the acceptance of any risk in the performance of his flying duties.

It is interesting to note that my friend owned or flew nearly every general aviation aircraft made after 1935. He was a CFI and was an airline pilot for 38 years -- DC-3 through B-747! I believe his 24 hour no drinking rule is well worth adopting.

The FAR's prohibit pilots from performing crew member duties within *"eight hours after drinking any alcoholic beverage or while under the influence of alcohol!"* Studies have shown that even after 8 hours a persons blood alcohol content may be high enough to cause some degree of impairment in the cockpit. Therefore, an excellent rule is to allow 24 hours between *"bottle and throttle."* Alcohol and flying are a potentially lethal combination. It is easy to participate in the social arena with a big glass of ice cubes and ginger ale. Try it!

MEDICATION

Pilots taking over-the-counter or prescription drugs should check with a physician, qualified to give FAA medical exams, to determine if any of these medications or combinations of medications may cause impairment. Very serious accidents have resulted when pilots were impaired by medication. Don't take chances -- check and be sure.

ILLNESS

Do not fly unless you are in normal good health. Do not fly with any symptoms of abnormality. The results are likely to be degraded physical and mental capability up to and including incapacitation. If you don't feel *"good"*, leave the airplane in the barn.

OXYGEN

It goes without saying that failure to use supplemental oxygen in accordance with FAR's or at any time good judgement dictates -- is dangerous. Smoking in flight may contribute to oxygen deprivation.

UNDERSTANDING COCKPIT RESOURCE MANAGEMENT

Can we practice Cockpit Resource Management when the crew is a single pilot? The answer to that question is a resounding *"YES!"* In the early 1980's United Airlines developed and implemented their Cockpit Resource Management (CRM) program under a different title. This involves all cockpit crew members on an initial and recurrent basis. Most airlines have now adopted the CRM concept under various titles. The results of the CRM training have been dramatic.

Single pilot CRM requires that we keep in mind that the captain, co-pilot and flight engineer functions are all wrapped up in **one person -- the single pilot!** This very sobering fact identifies the substantial responsibility of the single pilot in any and all types of operations. It also highlights the urgent need for an organized approach to decision-making. The key here is resource management. What are the resources available to the single pilot? How and when can these resources be utilized to make the best possible decision?

After the decision is made, how can these resources best be utilized to assist with a safe and orderly conclusion of the flight or in the original planning of a flight?

RESOURCES WHICH ARE AVAILABLE

Let's begin by listing the resources available **INSIDE** the cockpit:

1. Navigational maps/charts as appropriate
2. POH on board and readily accessible to the pilot
3. Normal and emergency check lists
4. Flight log and fuel burn-out log, complete and current at all times while in flight
5. Flashlight
6. Fire extinguisher
7. Copy of the flight plan
8. Copy of appropriate weather data for departure, enroute and destination
9. Specific airport data such as elevation, runway length, lighting, obstructions in the area and IFR approaches.
10. Passenger(s) - Watch for traffic, hold charts, etc.

The following resources are located **OUTSIDE** the cockpit and **readily available for the asking:**

1. Assistance from ATIS and Ground Control
2. Assistance from Clearance Delivery
3. Assistance from Tower and Departure Control
4. Assistance from Enroute ATC, including flight following
5. Assistance from Flight Watch and FSS while enroute
6. Assistance from Approach Control
7. Assistance via UNICOM
8. Assistance via 121.5 or air-to-air on 122.75
9. Assistance from airline/corporate and other aircraft via ATC center frequencies.

NOTE: These are 19 separate resources available to the pilot. There may be more. To be truly prepared, a pilot will always have at least the first 9 in-cockpit items on board. He may have additional items depending upon the type of operation being conducted. As for the 9 items *"outside the cockpit"*, they are all available **for the asking!**

Ask and you shall receive helpful assistance. The **BIG PROBLEM** is that we either, **don't ask for assistance in a timely manner or we never ask!**

Where to start? Pay strict attention to the ATIS when available. Play it over as necessary to be sure you have and understand the total message. If you don't understand the clearance from ground control, ask them to *"say again"*. If you have a special need, want a different runway, etc., say so! Controllers will assist you to taxi to the take-off runway, if you are not familiar with airport layout, or if at night or reduced visibility. I have found these folks to be very helpful, **but you've got to ask!**

The most important resource of all is a competent pilot (crew). To optimize crew performance, it is necessary to have good communications with everyone who can have even a small amount of input which may help in making a correct decision. Following the decision on what to do, comes the task of carrying out the procedure or operation to a safe conclusion. This task requires continuous monitoring for possible revisions. Do not *"sit on a problem."* Start corrective action promptly. Don't be afraid to declare an emergency. Many pilots feel a negative stigma will be attached to their piloting skills if they declare an emergency. This is ridiculous. In fact, a pilots savvy and judgement are lauded when a timely declaration of an emergency saves an aircraft or lives. Proceed in a logical and orderly manner to avoid errors which may exacerbate the problem at hand or lead you to miss identifying other problems. Key elements of CRM are well-directed investigation of the situation with input from all available resources thus leading to a skillful execution of the operational decision.

EXAMPLE: In the airline world, United Flight 232's emergency descent to Sioux City, Iowa, in August 1990 was a perfect example of excellent cockpit resource management. A catastrophic failure in #2 engine of the DC-10 ruptured hydraulic lines to all systems and left the rudder jammed out of center. The crew now had no operating flight controls! No aileron, elevator or rudder control. The aircraft began uncontrolled steep turning descent! Captain Al Haynes and his crew were not about to give up and burn a hole in the ground losing all lives on board. As a result of their heroic efforts, along with assistance from a deadheading DC-10 instructor and the tower controller at Sioux City, they got almost to the airport runway before the DC-10 crashed. Keep in mind that there was no written operating procedure for this situation. The expert application of CRM during this 30 plus minute ordeal enabled the crew to devise a method for minimal control of this huge aircraft. As a result 189 lives were saved!

EXAMPLE: The following is cockpit management by a single pilot. The pilot was on a VFR flight plan from Salt Lake City to his home in southern Colorado. The weather was forecast to be high ceilings enroute. West of Grand Junction he encountered engine trouble which reduced power so much that he needed to land soon. When he contacted the Grand Junction tower he learned that the local area was now under a cloud deck. The high surrounding terrain prevented getting into the valley VFR. He declared an emergency. The tower with cooperation from Denver ATC vectored him to the airport area where a descent was made to the downwind leg of the traffic pattern. The engine died on landing roll out. Excellent teamwork between pilot, tower and Denver ATC led to a safe landing with a sick engine! Great cockpit resource management. **Help is available if you communicate your needs!**

ANOTHER EXAMPLE: On occasion UNICOM can be a resource when used with a bit of ingenuity. For instance, during the winter of 1986, I was conducting a mountain cross-country training flight out of Denver. It was mid-afternoon when we landed at Leadville, Colorado. After a break of approximately thirty minutes, we took off and started south on course toward Buena Vista, planning east to Badger Mountain and back to Denver. Heavy snow showers were suddenly obscuring the mountain tops to the south. The route to Buena Vista was cut off completely. The only way out of the Arkansas River Valley was to proceed northbound from Leadville over Tennessee Pass then to Kremmling and via Corona Pass to the Denver area. But -- I can't contact the FSS from my deep valley location. I need to revise my flight plan quickly! I called Leadville UNICOM and asked them to call DENFSS on their direct line and give them my revised flight plan, time enroute, fuel on board and time of departure from Leadville. The trainee and I proceeded to Denver without incident, thanks to the FBO at Leadville.

CRM SUMMARY

By regularly utilizing all appropriate resources, INSIDE and OUTSIDE the cockpit, you can avoid many of the problems which cause accidents. In the rare instance where you encounter an unpredictable operational problem, you will be best prepared for a safe conclusion. Communicate, get assistance and do not hesitate to declare an emergency!

FAULTY ASSUMPTIONS

We need to guard against *"jumping to conclusions"* or making *"assumptions"* which are not warranted. **DO NOT ASSUME** anything. Always check in a logical manner to determine what is wrong or not wrong in the many operating situations which may be encountered.

When engine or systems abnormalities occur, go to the procedure in the POH or the check list. Do not assume that you know how to do every procedure from memory regardless of how long you've flown the same plane.

In the ATC arena never assume that because you are on an IFR flight plan you are being protected from a potential midair. See and avoid is necessary -- IFR clearance or not! Make a practice of being in the see and avoid mode anytime you can see beyond the wing tips.

EXAMPLE: In the late 1960's a Cessna 310 was on a training flight with a CFI and one student on board. On the landing approach the gear handle was moved to the down position but no green light was in view. The CFI alerted the tower of the situation and ordered the runway foamed. The gear was, in fact, down and locked! The only damage was a bill from the fire department for $700 for the foam. Yes, you do pay for the foam! The fact was that the gear had extended normally and the gear handle was left in the down position. The POH and check lists were on board but were never consulted because the CFI and the trainee *"assumed"* that the gear was *"jammed!"* A landing gear extended involves special noises and feels and slows the plane approximately 15 to 18 KTS. These items were not considered. A throttle could have been closed to see if the horn sounded, but none of these things were considered. Why? The problem was a failure in the wiring between the gear micro-switch and the light bulb for the green light. That's the best story I know about faulty assumptions. The use of the POH and check lists would have prevented a lot of stress, embarrassment and expense.

ANOTHER EXAMPLE: In Australia a sailplane pilot was on a long cross country when he was unable to get back to base because of weather conditions. He landed without damage to the sailplane and called friends at the home base to bring the enclosed sailplane trailer to haul his machine home. The friends hooked the trailer to their pickup and drove the 200 miles over poor roads to the landing site. The friends pulled up to the downed machine and opened the trailer doors. To their surprise there was a sailplane in the trailer from a previous retrieval! Now that is really a faulty assumption! These men **assumed** the trailer was empty when they began the trip--But it wasn't empty. A pre-departure check would have saved time, expense and embarrassment!

SITUATIONAL AWARENESS

This involves knowing what is going on around you at all times. Examples are:

Current status of the weather prior to departure and while enroute.

Amount of fuel on board at any given time. Not just the quantity but the *"TIME IN YOUR TANKS"*. Keep a running fuel log by time of fuel used from each tank. Do not depend upon the fuel gages.

The safe minimum altitude for each segment of the route including the immediate vicinity of the destination airport.

Ground speed being made good.

ETA to the next check point and the destination.

At mid-point of your cross-country flight make an estimate of fuel reserve expected at the start of your landing approach -- VFR or IFR.

Traffic awareness (collision avoidance) enroute and on departures and approach. Apply an orderly and continuous scanning technique.

Continually check the status of the aircraft, engine and systems operation.

Monitor the comfort and well-being of the passengers.

What alternate airports might be usable in case of a need to land for an medical emergency or a serious mechanical malfunction?

Always have an alternate airport in mind VFR or IFR. Unexpected strong crosswinds, a disabled aircraft on the runway, etc., may make your planned destination unusable. In such situations be prepared to proceed to the nearest suitable alternate airport.

SITUATIONAL AWARENESS CAN BE PRACTICED UNTIL IT BECOMES A STRONG ASSET. A PILOT WHO MAINTAINS A COMPLETE SITUATIONAL AWARENESS AT ALL TIMES IS TRULY A PROFESSIONAL WHETHER LOW-TIME STUDENT OR A HIGH-TIME AIRLINE CAPTAIN.

EXAMPLE: A light twin was landing after an IFR approach when the pilot realized that he had touched down long on a short runway. The pilot executed a go-around. On the following landing, the pilot forgot to lower the landing gear. After the props struck the runway, the pilot attempted still another go-around. The aircraft stalled and crashed off the end of the runway. The pilot was killed and the aircraft was destroyed by fire. Check lists are essential to good situational awareness.

EXAMPLE: In California a Beech Bonanza landed on top of a single engine Cessna which had just landed and was rolling down the runway. Both aircraft were heavily damaged and one pilot was seriously injured. The pilots were reported to be high-time! Situational awareness is important for every pilot. Traffic awareness can save lives -- yours and mine included.

ANTI-AUTHORITY ATTITUDE

This attitude has been observed among pilots for a very long period of time. Anti-authority attitude involves knowingly ignoring FAA regulations, airplane and engine operating procedures and weather conditions not appropriate to the operation being conducted. This attitude seems to hinge on erroneous understanding of vital operational information and/or a stubborn resistance to following generally accepted and legal operating practices. This individual seems to be saying, *"Don't tell me what to do. This is my airplane and I'll fly it as I damn please!" "The FAA consists of a bunch of dumb bureaucrats." "I've flown in ice and it's no big deal."* And so on. This type of individual may burn his fuel down to the fumes and boast about it. Ironically, pilots with an anti-authority attitude frequently *"get by with it."* They also tend to get involved in accidents of their own making -- many fatal! Their attitude and practices may influence other pilots to follow suit. Very bad!

EXAMPLE: In the late 1950's I was scheduled to instruct a local business man. He was extremely vocal and argumentative about everything I was attempting to impart. After about 45 minutes, I took control of the plane and terminated the training period. I explained that his attitude was going to get him killed and strongly suggested that he discontinue his quest for a pilot certificate. He scoffed at my advice. Less than a year later a rancher in the Texas Panhandle came upon the wreckage of a single engine Cessna. This man had departed on a night cross-country flight to Denver. Heavy thunderstorms had been forecast along the route and had, in fact, occurred. The pilot was dead in the wreckage. Enough said!

EXAMPLE: A pilot departed from a mountain airport in snow showers. Initial investigation pointed to a heavily overloaded cabin twin and no record of the pilot possessing a multi-engine rating. The airplane was totaled. The pilot and several passengers were killed.

EXAMPLE: A pilot flew into the ground on open grassland during a night cross-country enroute to his base of operations. The pilot was killed. The ceiling and visibility were very low at a reporting station some distance from the crash site. This individual had represented himself as having a commercial with flight instructor certificate when, according to FAA records, he didn't possess either of these certificates. This situation had gone unnoticed. He had arrived from out of state several years prior to the crash and no one had asked to see his pilot certificates.

EXAMPLE: There are pilots exercising the privileges of Private Pilot who have only Student certificates, and some exercising the privileges of a multi-engine rating when they do not have this rating. There are great risks beyond the obvious in not having the claimed FAA pilot certificate.

A sobering fact is that these misrepresentations void any insurance coverage on the airplane being flown. Read any hull and/or liability policy and you will note that any material misrepresentation of the *"facts"* (read that to mean falsification regarding pilot certificates or ratings) will be grounds for denying claims. It has been said that the insurance companies cause compliance with the FAA regulations. In any event, making false statements on an aircraft insurance application can lead to the policy being declared void by the insurance company. These are key examples of anti-authority behavior of the worst kind.

SUMMARY

There is more to the job of piloting than merely flying the airplane. There is a **management component** to the job which has been recognized in airline and military operations for many years. However, only recently has management crept into the world of the general aviation pilot and then in bits and pieces. If you truly utilize CRM principles you will possess a professional management capability for all your operations -- on the ground and in flight. See FIGURE 1-1 for the ideal allocation of time and attention for the skilled pilot -- old pro or beginner!

Make a practice of utilizing all available resources, as appropriate, from the time of planning until the aircraft is secured at the end of the flight.

Always proceed methodically in every action you take and apply professional good judgement throughout the PIC area of responsibility.

Be a manager/pilot in the most professional sense. It is time we ceased killing and injuring ourselves and our passengers by non-professional attitudes and practices.

THE DIVISION OF TIME AND ATTENTION FOR THE PROFICIENT PILOT

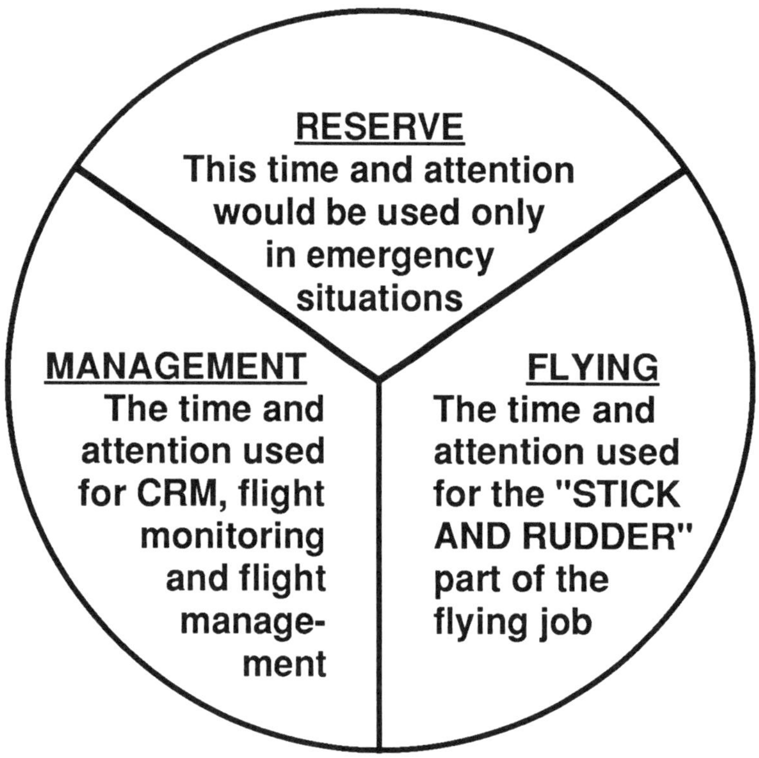

There is something wrong if the pilot requires more than 2/3 of his or her time and attention to carry out normal flying operations.

Proper understanding and utilization of CRM and good flight management will usually allow a good average pilot to comply with the 2/3 rule displayed here.

Figure 1-1

THE PRE-FLIGHT INSPECTION

TIE-DOWNS

If a tied-down airplane is to be truly secure against gusts and/or sustained high winds, both wings and the tail must be securely tied down.

The tail tie-down is very important in preventing the airplane from twisting from a strong crosswind component. This twisting is often severe enough to break the wing tie-downs or the tie-down attachments under the wings. Additional damage results when the plane turns thus striking adjoining planes. This is no way to become popular with your ramp neighbor or insurance adjuster.

Flight controls must be secured so that a tail wind will not cause the ailerons and rudder to be banged from stop to stop. Severe flight control damage and/or failure can result when the controls are not secured.

FROST, SNOW AND ICE

Flight must not be attempted with **any** frost, snow or ice on the prop or an airfoil. These items cause greatly increased drag and dangerously disturbed airflow over the wings and other flight controls surfaces. The net result is greatly reduced lift and reduced effectivity of flight controls and unpredictable stall characteristics.

THE SCARY FACT IS THAT ANY FROST, SNOW OR ICE ON THE AIRFRAME IS VERY DANGEROUS. JUST A *"LITTLE BIT"* HAS CAUSED COUNTLESS ACCIDENTS AND FATALITIES. DO NOT TRAP YOURSELF IN THIS DEADLY SCENARIO. ANY FROST, SNOW OR ICE ON THE AIRFRAME **DEMANDS A NO-GO DECISION** UNTIL REMOVAL IS 100 PERCENT COMPLETE.

Always make a visual check of the fuel quantity in each tank and compare with the fuel quantity indicators. Be sure fuel tank caps are secure -- a loose cap or one dislodged in flight will result in fuel being siphoned out at an awesome rate.

A note on fuel quantity -- *"Full"* and/or *"Topped Tanks"* will seldom provide the usable fuel quantity listed in the aircraft owner's manual or the fuel selector placard. The reasons for this are that the certification process involves a perfectly level ramp surface and a specified nose strut extension, plus the fuel filled until the tanks will hold no more. You should routinely expect to be short of a full fuel load by two to five gallons, depending on the capacity of each tank and the make / model of aircraft.

Drain the sumps and carefully check the fuel samples for color, smell and the absence of water or other contaminates. Keep in mind that water may be in your tank as a result of having been pumped in inadvertently during fueling or by condensation. I know of an instance where a light twin was fueled and ended up with over five gallons of water in each tank! Not a happy event. This pilot wisely drained both tanks completely and started over with fuel from another vendor. Check a fuel sample after each refueling.

Check the oil quantity with care. Many dipsticks are difficult to read, especially with clean oil. If you are not certain of the oil quantity reading, repeat the dipstick check -- do not guess! Know the minimum oil quantity set out in the POH. Be sure that the oil fill cap and/or dip stick are secure.

Check tire pressures to the prescribed amount. Low tire pressures allow excessive sidewall flexing during taxi, takeoff and landing. Sidewall flexing generates heat and leads to weakening or failure of the tire carcass. A blown or flat tire on takeoff or landing will provide you with more excitement than you care for.

LANDING GEAR AND BRAKES

Check the nose gear and the firewall for any signs of damage. Most nose gear assemblies are mounted to the firewall via the engine mount. A wrinkle (NOT THE STIFFENING GROOVES) in the firewall usually indicates that the nose gear has been over-stressed by a badly porpoised landing. Firewall replacement is a large job requiring engine and nose gear removal, rivet removal and replacement -- thus lots of labor. Have you priced a firewall lately? I recently had a call from a gentleman who has owned an older Cessna 182 for several years. He is a good and careful pilot. The firewall was bent so much that he noticed the damage while looking forward of the rudder pedals. He had not been aware of the firewall check until I mentioned it! But -- what about those past annual inspections? This was a very old problem which had gone undetected for a long time.

Check the main gear mounting area. Any wrinkled fuselage belly skin (Cessna) in the vicinity of the gear box may indicate structural damage usually caused by a very hard landing. On low wing airplanes (Beech, Piper, etc.), check for wrinkles on the bottom wing skin where the strut attaches to the airframe. Again, skin damage usually results from a hard landing. In a low wing airplane wing spar damage is also possible. If you have any doubt about suspicious structure around any gear mounting area, have it checked out by an A&P.

By the same token if you have had a hard landing, have the airplane checked before it is flown again. This is standard procedure with the airlines and for good reason. A good guideline to keep in mind is that when you hit the landing surface at the equivalent of 1000 to 1200 FPM descent or more, you can expect to do big damage. If you want to see the airframe *"TOTALLED"* try around 1500 FPM on touchdown! Those gear struts sticking up through the top skins on a low wing airplane, or bent straight out on a high wing airplane, are not a pretty sight!

Gear struts must have no fluid leaks and be properly inflated. Check the drag link *"scissors"* to be sure there are no cracks or evidence of excessive wear. If a drag link breaks during taxi, take-off or rollout, the wheel will be free to turn in any direction. Further, if a drag link fails while airborne, the strut may drop out of the mounting assembly and depart the aircraft. If the plane has been moved by a tug, make a detailed inspection to determine if the nosewheel has been forced beyond its R/L turn limits. If there is any doubt, have an A&P check this before operating the aircraft.

Gear doors must be checked to ascertain that they are secure, in proper alignment and not damaged.

Brakes must be checked for excessive wear, fluid leaks, etc. Experiencing a complete brake failure during taxi is doing it the hard way -- very scary and hazardous. Check that you have brake pedal pressure before you start the engine. This may prevent a nasty surprise if you have no brakes after the aircraft has begun to move. Make a second check of the brakes as you begin to taxi.

COWL FLAPS AND EXHAUST SYSTEM

Cowl flaps should be checked for security and for excessive wear of the hinges or operating controls. Cowl flaps are subject to heavy vibration when open in flight. In some severe situations a cowl flap has come loose in flight; held only by the control rod while it furiously banged against the belly of the plane.

EXHAUST STACKS

Exhaust stacks must be checked for security. If there is any rattle or excessive movement of the exhaust stack, you may have a partially burned through stack, loose stack or a muffler about to fail. These situations may result in a fire in the engine compartment or carbon monoxide (CO_2) in the cabin. Insist on making a thorough check of the exhaust system. Several years ago a pilot took off in his light twin and a few minutes later reported a severe fire in the right engine. He tried to get back to the airport but crashed and burned. Back at the ramp was the exhaust stack from the right engine, it had fallen off not far from where he began to taxi!

ENGINE COMPARTMENT AND COWLING

Always check the engine compartment for any loose connections, fuel or oil leaks and stains. Also check for a birds nest and assorted debris. The odor and fire hazard of an overheated or burning birds nest, in flight, in unpleasant at best.

Engine cowling must be secure before starting the engine. If in doubt, open the cowl again and start over. Do not force a cowling closed! Be sure it is securely fastened. If an engine cowling which is hinged at the top comes loose in flight, you should expect it to come open to about the same position as when you are checking the oil.

The cowling may tear loose and depart the aircraft with a crash into the windshield and/or over the top of the fuselage. A one piece top cowling needs to be secure, otherwise its departure in flight will cause great damage and could injure the pilot and front seat passenger.

I still have a vivid 1952 memory of the left top cowl on a Cessna 140 coming unlatched in flight. The cowling fluttered wildly in a nearly upright position in front of my unnerved student. I immediately reduced RPM and slowed to approach speed. We were successful in returning to the airport before the cowl hinge assembly failed.

PROPELLER

Check the leading edges for nicks and scratches. The first fatal aircraft accident in the U.S. was caused by a propeller blade failure. The time was September 17, 1908. The pilot was Orville Wright who was seriously injured; however, his on-board military observer was killed. The smallest nick or scratch becomes a stress point which may be the location for separation of part of the blade, a deadly situation. If in doubt, have an A&P check the prop and work out the nicks. On constant speed props check each blade for security of the mounting in the hub by noting that a blade will not rotate in the hub. There have been several instances where one blade rotated in the hub to a near feathered position while the other blade remained in the commanded pitch setting. This will result in inability to continue flight.

During a recent two-year period 83 accidents occurred due to propeller blade failure. There were 21 fatalities. These blade failures were sudden and catastrophic. The rate at which a nick/crack separation failure progresses is not predictable.

If your propeller has nicks, cracks or scratches, it is a *"sick"* prop. Get it fixed before you fly! Bottom line -- should you lose a part of blade in flight, **IMMEDIATELY** close the throttle, that is, if you can see it and get your hand on it! You should expect the instrument panel to be a wildly vibrating BLUR!

AIRFOILS AND FLIGHT CONTROLS

Check flight controls thru full travel. When inspecting wings, horizontal stabilizer, vertical fin and rudder, look for wrinkles in the skin. Wrinkles may indicate that the structure is bent or broken and unsafe for flight. Cracks should be stop-drilled. Check ailerons, elevator, rudder and all trim tabs for movement in the proper direction in response to operation of the cockpit controls. Each year we have accidents, usually fatal, where flight controls are connected the reverse of normal.

A rebuilt cabin twin being test flown by a well-qualified pilot crashed on initial climbout killing the pilot. The aileron controls were found to be reversed! I had the experience of the elevator trim tab control being connected backwards. When my low-time student reached 1000 feet AGL and applied nose down time, the nose rose abruptly! This situation could have had a different resolution had the pilot been a low-time solo student.

These flight control problems are most likely to occur when flight controls have been disconnected for painting the aircraft, major airframe repair, or replacement of flight controls, cables, push rods, etc. Control cables have come off pulleys. Pulleys have frozen and no longer turn, push rods have become disconnected, etc. Be in the role of a *"nit picking"* test pilot every time you conduct a preflight!

In 1953 I was a charter pilot flying a Lockheed 10A (shades of Amelia Earhart). This plane had a hard life during and after WWII. When it went in for a check at Chicago, I mentioned that the elevator control seemed a little stiff. In the far aft fuselage a pulley carrying one of the elevator cables was frozen in place and didn't turn. This had resulted in the control cable being nearly worn through. The A&P stated that the elevator cable was very close to separating -- not a fun prospect, especially during a takeoff or landing. Check the flaps for security and operation.

MISCELLANEOUS ITEMS

Check all exterior lights, pitot heat, pitot static system, fuel tank caps and vents, air intakes and filters and the windshield.

AS A FINAL CHECK, WALK AROUND THE PLANE SLOWLY AT A DISTANCE OF 15 TO 20 FEET, STOPPING DIRECTLY IN FRONT OF AND DIRECTLY BEHIND THE PLANE. THIS FINAL CHECK WILL AFFORD YOU AN OPPORTUNITY TO SEE ANY DEFORMITY IN WINGS, TAIL, FUSELAGE, ETC., PLUS GEAR STRUT AND TIRE INFLATION.

A Cessna 200 series airplane had been damaged and subsequently flown a few times. Standing behind the plane, a slight bend in one side of the horizontal stabilizer could be seen! The stabilizer spar was cracked. The stabilizer had to be disassembled and a new spar installed. The punch line is that the plane had been flown several hours with the damage undetected! Look carefully and look again! If you ever believe you may have damaged something, report it and have an A&P check it. Do not take any chances on causing an accident that may occur on a subsequent flight.

FINAL NOTE

If your plane is hangared, the preflight must be just as thorough as in the case of a tied-down aircraft. Remember, the most unexpected things happen when airplanes are NOT FLYING! **Be on the alert for damage from** *"HANGAR RASH"* **which may make you plane unsafe to fly!**

Carefully check the gross weight and check that the loading is within CG limits. **CHEATING CAN GET YOU KILLED!** Secure all cargo and baggage.

The foregoing items are worthy of special attention because they are associated with many accidents.

Your Owners Manual or POH furnished by the manufacturer should always be followed in conducting the preflight inspection. Remember, the FAA-approved manual is a legal document. A safe flight always begins with a thorough and thoughtful preflight inspection. The preflight check list in the owners manual should be considered to be the **MINIMUM** in conducting a preflight inspection. Always assure that your preflight is thorough and complete for your plane and your planned flight. Do more than the minimum.

The lack of good preflight inspections cause all sorts of *"bumps in flight"*. Fuel quantity and fuel management problems are prominent, but there seems to be no limit to the surprises. Read on.

COULD THESE ACCIDENTS HAVE BEEN PREVENTED?

A light twin departed a New Jersey airport and a right engine fire broke out during climb. The pilot returned to the airport where the fire was extinguished. The fire resulted from the outboard main fuel line chafing against the exhaust manifold. The line had been improperly installed.

A Cessna 150 was landing at an Arizona airport when on touchdown the airplane veered left and departed the runway. The nut securing the nosewheel had backed off allowing the nosewheel to become misaligned.

A single engine Cessna departed a Nebraska airport and after about an hour, the engine began to lose power. During the forced landing, the aircraft was substantially damaged. Investigation revealed that the fuel was contaminated with water. The pilot was faulted for not having conducted an adequate preflight inspection.

The pilot of a Piper PA-24 reported that he ordered the aircraft serviced with fuel during a stop at a southern airport. About an hour after takeoff, the engine lost power. The pilot then switched tanks but was unable to maintain engine power. During the ensuing forced landing, the airplane was heavily damaged. The pilot acknowledged that he did not visually check the fuel quantity before departure.

A Champion 7GBC departed Jackpot, Nevada, on a cross-country flight to Idaho. During a left turn, the pilot experienced an uncommanded left roll but was able to level the wings just before crashing into rough terrain. Investigation revealed that an aileron attach bolt was missing.

A Beech Bonanza departed a Florida airport with an immediate power loss. The airplane stalled and crashed. Investigation revealed water contamination of the fuel system.

A Cessna 182 was on a cross-country flight in Michigan when the RPM dropped to 900. The pilot applied carburetor heat but no improvement so he carried out a precautionary landing. Water and debris were found in the fuel strainer and carburetor bowl.

At a Rocky Mountain airport a takeoff was begun but the airplane barely staggered into the air then crashed into a river. Frost on the wings is suspected as the cause of the crash.

A cabin twin crashed just past the end of the runway on takeoff. NTSB/FAA is checking the possibility of being over gross weight, ice/snow on the wings and engine malfunction.

> NOTE: DURING THE PREFLIGHT INSPECTION -- DO NOT FALL FOR THE OLD DEMONS *"GET THERE ITIS"* OR *"GET HOME ITIS"*. THESE MAKE US RUSH OR NEGLECT OUR PREFLIGHT ACTIVITIES AND PLANNING. NOTHING YOU *"NEED TO DO"* TOMORROW IS WORTH RISKING YOUR LIFE AND OTHERS TODAY. EXERCISE PROFESSIONAL LEVEL RESPONSIBILITY!

WEATHER BRIEFING

The questions to be answered are - How do I get a good weather briefing? What information is needed if the briefing is to be fully adequate for the flight being planned? A thorough weather briefing is important for two reasons. First -- the go - no go decision usually hinges on weather conditions. Unless the weather briefing is complete and fully understood, the pilot is likely to make a flawed go-no go decision. Second -- Get a standard briefing with special attention to winds aloft, terminal forecasts, local and enroute weather sequences, sigmets, NOTAMS, etc. Ask questions as necessary in order to acquire all the information appropriate to your flight. Now let us turn to things that are easy to do which will greatly enhance the understanding of the overall weather situation for the planned route and time of flight. The following is a list of additional resources for weather information:

1. A. M. WEATHER if available
2. THE WEATHER CHANNEL if available
3. LOCAL TV STATION weather maps/reports
4. NEWSPAPER weather maps
5. HI-TECH weather services

Each of the above resources provide a valuable snapshot of the overall weather situation at the present and for several hours into the future. Combining the information from these resources with a thorough FAA weather briefing will provide an optimum understanding of the weather.

Regardless of how you receive the FAA weather briefing information -- the following are significant items worthy of special consideration.

FRONTS

What are the locations and types of fronts? What is the direction and speed of movement? What are the weather characteristics during station passage of the front or fronts in your area or route?

TEMPERATURE AND DEW POINT

If these are not given -- ask for them and record them. These are a basis for forecasting fog, ceilings, low visibility and precipitation. A trend of temperatures and dew points close together, (within 3 degrees F), over a widespread area is a clue to what the airmass is doing or is likely to do along your route.

When will the surface temperature be highest during a typical day? During winter months the temperature peaks at approximately 12:00 noon and begins to decrease significantly after 2:00 P.M. During summer months the temperature peaks at approximately 2:30 P.M. and decreases significantly after 5:00 P.M. For example -- if a person plans a summer departure at 09:30 A.M., and the destination weather is at minimums for the operation, much improved weather may be the case upon arrival at 1:00 P.M. On the other hand, a 4:00 P.M. departure for 7:30 P.M. destination arrival may find weather which had deteriorated significantly. These examples, may of course, be shot down when frontal passages occur along the route of flight.

Is weather forecasting an exact science? The answer is **NO!** Weather forecasting is erroneously accepted as "*gospel*" by pilots when in fact the forecasting is generalized to a great extent. I suggest that the responsible pilot must utilize additional resources, including his personal knowledge of weather, in order to adequately prepare for operations in marginal weather conditions.

PILOT REPORTS

Pilot reports are valuable and should be sought out and utilized prior to departure and while enroute. These have special usefulness because they describe the actual conditions encountered by fellow pilots.

Carefully consider the time of the pilot report, the location and altitude of the in-flight observation. Obviously the most reliable pilot report is one made along your route of flight, at or near your actual or proposed altitude and no more than 1 hour old. The greater the deviation from these ideal guidelines the less the value of the pilot report. **Keep in mind that weather conditions are in a constant state of change!** It then follows that we need all the help and information we can muster in order to operate safely and reliably.

UPGRADE YOUR WEATHER KNOWLEDGE

Study and review the FAA publication "*AVIATION WEATHER*" which is available at pilot supply shops or from the Superintendent of Documents, U.S. Government Printing Office, Washington D.C. 20402. I recommend a thorough review of Part 1, which is chapters 1 thru 12. The book entitled "*WEATHER FLYING*" by Captain Robert Buck explains how weather develops and how it is flown in with safety. The author is a retired airline captain and general aviation pilot with several decades of experience in every imaginable weather situation. His writings are directly applicable to general aviation operations and in my opinion this is the best of the aviation weather books. This book is easy, fun reading and very educational because of the real life experiences portrayed. This book is available from **SPORTY'S PILOT SHOP** 1-800-543-8633.

FLIGHT PLANNING

When a winds aloft forecast states a headwind of 30 KTS, or more you should prepare to encounter a much stronger headwind aloft. The strategy I suggest is to increase the fuel reserve by 15 minutes of burn out for each 1 hour of planned flight time.

For example, if you plan a 3 hour flight in this condition add an additional 45 minutes of reserve fuel. The alternative is to plan a contingency fuel stop enroute which can be utilized in case of stronger than forecast headwinds. **Running low on fuel is inexcusable** -- and is to be avoided.

THE FLIGHT PLAN AND FLIGHT LOG

The flight plan may be high tech or a simple paper and pencil document. In preparing a flight plan keep each leg of the flight at a maximum of 3 to 3 1/2 hours. It is my experience that longer flight legs leave the pilot fatigued physically and mentally. To bring this point into focus--try sitting on a straight chair continuously for 4 hours or longer! Corporate and airline crews can alternate at the piloting job and get out of their seats and stretch periodically. Additionally, you need to give the passengers a break. Do not arrive at your destination to conduct a VFR night or IFR approach while physically and mentally fatigued -- it is foolish and dangerous. Look at the percentage of accidents which occur during approaches and landings after long cross-country flights.

The route of flight and method of navigation should involve what best meets the needs of the pilot in conducting a safe and orderly operation. Pilots with instrument ratings should routinely file IFR and operate accordingly.

This practice keeps the pilot *"in the system"* thus maintaining operating efficiency. Direct routings may be bothersome or unworkable in some areas because of airspace restrictions. Examples of heavy restrictions are the Eastern U.S., the Gulf Coast and West Coast. Thus, in VFR operations I prefer to proceed via airways when practicable. For VFR operations flight following is strongly recommended.

Additionally, when operating VFR, contact enroute approach control/towers when operating near or thru their airspace regardless of altitude.

A flight log should be maintained so that your position, ground speed and fuel state can be quickly determined at any time while enroute. I've had generators and alternators fail, batteries go dead in flight, voltage regulators malfunction, electrical wiring smoke and burn, circuit breakers and fuses fail and all sorts of failures of navigational equipment plus engine failures. It then becomes obvious that we need to know our geographical position at all times while enroute. I recommend a clipboard on the pilot's lap or readily accessible which holds the flight log, a copy of the flight plan, the weather information, plus the appropriate navigational charts and approach plates. Having these items at hand allows for a very professional operation at all times including times when you encounter unexpected failure of aircraft systems, an engine or navigational equipment. Keep in mind that new and expensive equipment can and does fail in flight. The bottom line is that the pilot is responsible for the safe operation of the aircraft at all times -- including after malfunctions or emergencies in flight. When the going gets tough the pilot needs to have the aircraft POH, check lists and a flight log at hand in order to operate in a safe and orderly manner.

AIRCRAFT LOADING

It is absolutely necessary that the pilot understand and utilize the weight and balance data for the aircraft being operated. It follows that the maximum gross weight and C.G. limits must be checked and complied with prior to each flight. In case of the forward C.G. being out of limits the pilot may not have sufficient elevator authority to lift off and/or flare to land. In the case of aft C.G. being out of limits the aircraft may lift off prematurely and/or pitch up uncontrollably on landing or takeoff. In either case, the aircraft may be uncontrollable at lower airspeeds including the approach and landing phase. Additionally should the airplane stall and/or spin, with the C.G. beyond the aft limit, recovery will likely be impossible. Always comply with the weight and C.G. limits established by the manufacturer and the FAA during the certification process. Beware of situations where fuel burn causes the C.G. to move aft while enroute. The C.G. must then be calculated for the landing as well as for the takeoff.

IT IS NOT UNCOMMON FOR HANGAR FLYING SESSIONS TO INCLUDE PILOTS BOASTING ABOUT OVERLOADED/OUT OF C.G. OPERATIONS. I STRONGLY RECOMMEND THAT THESE STORIES BE IGNORED. FIRST OFF, THEY ARE PROBABLY UNTRUE, AND SECONDLY IF THEY ARE TRUE THEY REPRESENT THE ATTITUDES AND PRACTICES OF VERY STUPID AVIATORS.

COULD THESE ACCIDENTS HAVE BEEN PREVENTED?

A 690A Aero Commander was enroute from Dallas to Oklahoma City when it crashed. Airframe icing was listed as a factor in the accident. During the weather briefing, the pilot was not advised of a forecast for significant icing in Oklahoma which was contained in the area forecast issued by the National Weather Service.

A Beech Baron crashed while enroute form Teterboro, New Jersey to Saranac, New York. The weather briefer had failed to pass to the pilot a forecast for occasional moderate turbulence and light occasional moderate icing which was contained in the area forecast issued by the National Weather Service and pertinent to the route of flight.

In Oregon, a private pilot flying a Cessna 180, with two passengers on board, crashed at the top of a ridge line and all were killed. The plane was observed flying toward a pass at low altitude near the base of low stratus clouds. Probable causes were listed as failure to get a weather briefing, inadequate pre-flight planning and preparation plus continued VFR flight into IMC.

In Alaska, a commuter aircraft crashed short of the runway during an approach. The airplane was alleged to be over gross weight and with the C.G. beyond the aft limit.

A pilot operating a Cessna T-210 ran out of fuel in the traffic pattern at his home airport. This pilot had flown a round robin business trip over this route many times before the accident. Where was the planning? It is very important to know how much fuel has been burned and how much remains on board as reserve. The aircraft was severely damaged but there were no serious injuries.

NTSB investigations reveal hundreds of accidents similar to the ones covered here. No one likes to be involved in an aircraft accident -- so why do they continue to occur? To prevent these accidents each and every pilot must get serious about planning the flight before departure -- then monitoring the flight continuously while enroute.

In summary we need to concentrate on avoiding VFR operations into IMC conditions. Know the weather enroute and avoid running out of fuel. Know how much fuel has been burned. Plan to keep a proper fuel reserve at all times! If we could avoid the trap of these two scenarios we would make a significant decrease in accidents, injuries and fatalities.

PLANNING A FLIGHT SHOULD BE TAKEN JUST AS SERIOUSLY AS FLYING THE FLIGHT. THE MERE FILING OF A FLIGHT PLAN DOES NOT GUARANTEE THAT THE FLIGHT HAS BEEN PLANNED IN ALL ASPECTS! HUNDREDS OF ACCIDENTS WERE SPAWNED BEFORE THE AIRCRAFT LEFT THE GROUND -- BECAUSE OVERALL PLANNING WAS INADEQUATE, UNREALISTIC OR NON-EXISTENT. PLAN! PLAN! PLAN! BE A RESPONSIBLE PILOT IN COMMAND.

THE CHECKLIST DILEMMA

Most checklists for general aviation aircraft are now up to 30 years old and may need to be expanded and updated to accommodate current on board equipment and operational requirements. Keep in mind that nothing may be deleted from the POH. The goal is checklist enhancement!

This dilemma can be solved by duplicating the normal operating procedures from the POH, then make a rough draft by adding items as appropriate. When a complete draft checklist is at hand a final product can be printed. I strongly recommend a HAND HELD checklist! A good option is a HAND HELD electronic check list. Many pilots use G-U-M-P or any of several other memory check lists. It is my opinion that memory checklists are of little value. Most of these schemes fail when needed most urgently -- during abnormal or emergency operations and when the pilot is fatigued or distracted.

Emergency procedures checklists need not be revised. It is very important that the POH be on board and readily accessible to the pilot at all times on the ground and in flight.

SUGGESTED CHECKLIST ENHANCEMENTS

PREFLIGHT CHECKLIST

Add the following items to be accomplished as the first items on the preflight checklist.

COCKPIT CHECKLIST (* IF RETRACTABLE GEAR)

* Gear handle - DOWN
\- Elect. Master SW.- ON
* Gear Lights - GREEN ON (1) or (3) as required
* (If installed) Mechanical Indicators - check for gear down position.
\- Fuel gauges-CHECK QUANTITY--COMPARE WITH VISUAL CHECK
\- Elect. Master SW.- OFF
\- Required documents - ON BOARD
\- POH - READILY ACCESSIBLE TO THE PILOT

ENGINE START CHECKLIST

Add any or all the following items to be accomplished as soon as the engine starts:

OIL PRESSURE - INDICATING (WITHIN 30 SECONDS OR SHUT DOWN ENGINE)
RPM - SET 1000 RPM MAX
AMMETER - IND. CHARGE

AFTER START AND TAXI CHECKLIST

Add any or all the following items as appropriate.

- PITOT HEAT - ON - if O.A.T. is 40 degrees F or less and any moisture present or anticipated.
- FLAPS - CK UP
- BREAKAWAY POWER - 1200 TO 1400 RPM (APPROX.)
- BRAKES - CHECK
- TAXI RPM - 1000 TO 1100 RPM (APPROX.)

BEFORE TAKEOFF CHECKLIST

Add these items as appropriate.

PILOT HEAT - ON - If O.A.T. is 40 degrees F or less and any moisture is present and/or at any O.A.T. if IMC may be encountered during the climb to cruise altitude.

- DIRECTIONAL GYRO - CHECK AND SET
- TRANSPONDER - ON and ALT.
- TAKEOFF AND INITIAL CLIMB PROFILE - REVIEW
- POWER CHECK - VERIFY AFTER TAKEOFF POWER APPLIED
- ROTATE SPEED (See POH) _______ KTS/MPH
- CLIMB PITCH ATTITUDE _______ DEGREES N.U.
- INITIAL CLIMB SPEED (V_y) _______ KTS/MPH
- EXPECTED RATE OF CL. _______ FPM (APPROX.)

NOTES ON PITOT HEAT

The airlines turn PITOT HEAT - ON after engines are started and leave it on until parked at the end of the flight. Most general aviation pilots hesitate to turn PITOT HEAT - ON until after they encounter icing conditions.

Think PITOT HEAT - ON as being preventative action to be taken any time there is the slightest possibility of need. The PITOT HEAT is a part of the airplane equipment and failure to utilize it early enough has caused the death of too many pilots and passengers.

CLIMB CHECKLIST

Add this item.

DIRECTIONAL GYRO - RECHECK

NOTE: On some aircraft the gyro and vacuum pump system requires a few minutes of high engine RPM to stabilize. For this reason the directional gyro may need another adjustment shortly after departure.

CRUISE CHECKLIST

Additions to the cruise checklist are rarely needed.

APPROACH CHECKLIST (A NEW CHECKLIST)

(When 25 miles or more from landing or 10 miles prior to traffic pattern entry if local VFR)

- SEAT BELTS/HARNESS- SECURE
- ATIS - ACQUIRE
- FUEL ON BOARD - COMPUTE (FROM FUEL LOG)
- FUEL SELECTOR - SET FOR LANDING

Having the above items accomplished well in advance of reaching the terminal or traffic pattern area will minimize distractions when the pilot workload is heaviest. This will allow for an orderly operation IFR or VFR.

BEFORE LANDING CHECKLIST

You may want to enhance your checklist by adding the following:

1. Insert the appropriate IAS for each flap setting you may use during approaches. These numbers should be 1.3 X V_{so} for the weight and flap setting used.
2. Add the GO AROUND/MISSED APPROACH checklist immediately following the AFTER LANDING checklist so that it can be reviewed as appropriate.

GO AROUND/MISSED APP. CHECKLIST

You may want to make a checklist as follows:

- POWER- MAX T/O
- PITCH- SHALLOW CLIMB
- AIRSPEED- CHECK
- FLAPS- UP TO ABOUT 1/2 THEN UP TO ZERO
- GEAR- UP
- PITCH- ADJ. TO V_Y IAS (ABOUT 8 - 10 DEGREES NOSE UP)
- COWL FLAPS- OPEN

NOTE: Check the POH for your aircraft.

SHUTDOWN CHECKLIST

Add the following item.

ELEVATOR TRIM - TAKEOFF RANGE

This is a long standing airline checklist item which may prevent an unexpected and abrupt nose pitch up on the next takeoff in case the item is missed on the takeoff checklist.

SUMMARY

A pilot's thorough and consistent use of checklists is a mark of a true professional. The amount of flying time, the FAA certificates/ratings held and size of the aircraft do not in any way excuse the pilot who fails to use all appropriate check lists while operating an aircraft!

A printed and hand held check list is recommended. These checklists may be on paper stock or cards and possibly laminated. There are electronic checklists available which may fit your needs. Bottom line -- have good checklists and use them consistently -- C-150 to B-747!

The best method of checklist use is to develop cockpit patterns for operating or setting controls and switches. Follow on by running the checklist to pick up any items which may have been missed on the *"set up"*. Consistent checklist discipline is a mark of a good and safe pilot.

ENGINE START

Over the years I have noticed frequent abuse of aircraft engines with excessive RPM at engine start. Professional use of engine start procedures and careful attention to the tachometer(s) will minimize engine wear and maintenance expense. The following procedures and techniques have worked well for me and I recommend them to all my fellow pilots.

GENTLEMEN/GENTLEWOMEN - Start your engines! Let's emphasize *"GENTLE"*. Up to 50% of engine wear occurs during and immediately after engine start! Be sure the throttle setting complies with the POH procedure. When the engine starts -- immediately adjust the throttle to 1000 RPM or slightly less. We know that oil pressure does not build instantly. Further, the initial oil pressure indication does not mean that all engine parts are being lubricated normally. The engine(s) should be run at 1000 RPM or less for approximately 2 minutes before beginning to taxi.

When the engine is started at an O.A.T. of 60 degrees F or less -- or if the engine has been shut down for more than 1:00 -- the RPM will increase slightly after being initially set at 1000 RPM. This is the result of the lubricant being warmed by friction. Re-adjust the throttle(s) to maintain 1000 RPM or less. One or two throttle adjustments will usually suffice.

Hot Starts with fuel injection engines is a subject which spawns a variety of favorite procedures. Keep in mind that there are more than one manufacturer of fuel injection systems. **They are not all alike!** Thus it is mandatory that the pilot follow the regular start and hot start procedures as set out in the aircraft POH. Each engine seems to have a life of its own, when it comes to hot starts.

With an abundance of preferred personal hot start procedures we have an exciting subject for discussion among pilots.

Cold weather starts -- If the O.A.T is lower than +20 degrees F the engine(s) should be pre-heated before starting. Engine start at +10 degrees F or lower without pre-heat are difficult if not impossible. If a cold engine start must be done without pre-heat it is advisable that the oil be drained, warmed and refilled just before the engine start. Check the POH and proceed with caution. A major concern during any cold engine start is to prime the engine properly before engaging the starter. Have the throttle set properly then engage the starter. When the engine fires -- **DO NOT MOVE THE THROTTLE** until the engine has run several seconds. Do not *"PUMP THE THROTTLE"* because to do so may set the stage for an engine fire in a carburetor engine. In any case an abrupt opening of the throttle, immediately after the engine fires, will cause the mixture to be leaned so much that the engine will die immediately. Begin with the throttle set properly -- then leave it there until the engine has run a few seconds. When the throttle is increased it should be done gently and smoothly! Check the POH and if the procedure calls for *"pumping the throttle"*- do so, but if it doesn't call for this action, **DON'T DO IT!** Follow on with other appropriate steps to bring things to life - then go to the **AFTER START AND TAXIING CHECKLIST.**

TAXIING

Are you really ready to taxi? Look around carefully for chocks, vehicles, other planes and personnel. Be sure the area ahead and around the aircraft is clear. Apply breakaway power (1200-1400 RPM) and when the plane begins to move check the brakes.

Then resume taxiing utilizing approximately 1000 RPM with adjustments as needed. Strive to control ground speed by power alone insofar as practical and safe. Practice so that you can **TAXI BY THE TACH!** We use the tachometer in flight - so why not use it at all times such as engine start and during taxiing? Let's take advantage of the RPM indications to help us taxi more professionally. Taxiing is a very important operation and worthy of our careful attention. Unfortunately many pilots have not been taught to taxi and to use the brakes in a professional manner. Self practice using the guidelines set out here will make a pro of the most sloppy ground operator! Your passengers will applaud your smooth taxiing enroute to the runway and during the trek from the landing roll to the parking/hangar area at the end of the flight.

GIVE THOSE BRAKES A BREAK - Taxiing by the TACH and professional use of the brakes go hand in hand. Never apply the brakes unless the throttle(s) are closed except when turning, stops for engine run up or some emergency situation. A pilot utilizing RPM to regulate taxi speed will virtually eliminate the need to use brakes during taxi operations. Keep in mind that a 4 to 6 place airplane weighs approximately the same as a family automobile -- but has less than 20% if the brake surface area! General aviation aircraft brakes are by no means *"heavy duty"* assemblies! Brakes should always be used sparingly. Several years ago a major airline and its' pilots began a program of *"think ahead and save the brakes"*. The result was a 50% reduction in the cost of brake maintenance and smoother rides for the passengers. Every pilot should use the brakes like a pro.

WINTER OPERATIONS POSE ADDITIONAL PROBLEMS ASSOCIATED WITH TAXIING ON SNOW OR ICE COVERED RAMPS AND TAXI STRIPS. TAXIING ON SLIPPERY RAMPS ETC. IS TRICKY BUSINESS. USE EXTREME CAUTION, TAXI SLOWLY, USE BRAKES GENTLY AND ALLOW EXTRA CLEARANCE BETWEEN YOUR AIRCRAFT AND OTHER OBJECTS.

Winds in excess of a gentle breeze may blow the aircraft across an ice covered ramp or off the taxi strip out of the pilot's control. Slippery airport surfaces and strong winds are a prescription for deep trouble! Excessive brake application which locks the brakes will make an airplane into a sled in a heartbeat! When the wheels are skidding you are out of control. Keep in mind that there are many situations where taxiing on icy or snow covered surfaces can not be done safely and should not be attempted. Avoid taxiing through slush or water when the temperature is at or near freezing especially in low wing and/or retractable aircraft. Water or slush may be splashed or blown into flap, aileron or elevator controls and into gear wells where it can freeze. If it is necessary to taxi through water or slush, do so very slowly.

SELECTING A SPOT FOR THE RUNUP

During slippery conditions look for a dry spot which can usually be found even if it is on the parking ramp, a protected area by a hangar etc.

THESE INCIDENTS SHOULD NOT HAVE HAPPENED

In my early piloting experience at Kansas City -- my plane was blown about 50 feet across the ramp as I was taxiing on ice with a fairly strong crosswind. Luckily nothing was hit but this is a very helpless feeling and potentially dangerous!

During a fuel stop at a midwestern airport I watched the owner of a very expensive twin abuse his engines while starting them. Without checking outside the aircraft or calling *"clear"* he started #1 with an alarming roar, and at an RPM much closer to 1500 than 1000. He made no RPM adjustment to #1 while he started #2, which also accelerated to a very high RPM with no subsequent adjustments. At brake release the plane lurched forward and they were on their way! I wondered what his passengers thought of that rough and noisy start. How much had he shortened the life of the engines? Do engines take care of pilots the way pilots take care of engines? This is a sobering thought indeed!

COULD THESE ACCIDENTS HAVE BEEN PREVENTED?

In South Carolina when a pilot was unable to find assistance to remove the nose wheel chock from the aircraft, he applied enough power to jump the chock. He was unable to retard the throttle before colliding with a parked aircraft and striking a concrete wall! The aircraft was substantially damaged.

A Cessna 172 was taxiing out in winds gusting to 29 knots, while taxiing crosswind the plane was overturned. Airplanes with light wing loadings should be left in the barn during high wind conditions unless wing walkers can be provided for the ground operations -- and even then it is poor practice.

A Cessna 172G landed at Las Vegas, NV with wind reported as 24 knots gusting to 31 knots. The pilot exited the landing runway and while turning crosswind the airplane was flipped inverted.

A single engine 4 place plane was taxiing out for takeoff at Anchorage, Alaska International Airport.

The ground controller had instructed the pilot to taxi behind a taxiing Boeing 727. The pilot was issued a caution for jet blast. The jet blast from the B-727's engines caused the left wing of the 4 place airplane to be lifted into the air. Substantial damage was done as the aircraft came to rest on its' nose and one wing.

In the early 1980's a Japan Airlines B-747 at Anchorage, Alaska was taxiing for takeoff. The airport was reported to be a *"glare of ice"* with a very strong wind. As the aircraft turned crosswind on the taxi strip it was literally blown off the icy strip into the snow and mud.

During the early 1950's I made a fuel stop in eastern Kansas while flying a Jacobs powered Cessna 195. In my haste to start the engine I deviated from the normal priming procedure. When I engaged the starter the engine turned thru and backfired. When I released the starter I heard, at the bottom of the engine, what sounded like bacon being fried. I recognized this as a carburetor fire and immediately primed the engine properly and engaged the starter. The engine started and sucked the fire up into the intake! The outcome would have been different if my battery had been low, if there had been no fire extinguisher available, etc. When starting an engine follow the procedure carefully -- always.

SUMMARY

Engine starting and taxiing operations are extremely important because of the special hazards involved. It is here that we are closest to airport personnel, equipment, buildings and other aircraft.

Even under ideal weather conditions we are constantly at risk of collision with fuel trucks and other airplanes. Engine start and taxiing operations deserve the utmost professionalism and caution.

RUNUP

Prior to beginning the run up, the airplane should be aligned into the wind if possible. The surface of the run up area should be clear of sand, stones, and debris which could be pulled up into the prop(s) during the run up. Avoid stopping on surfaces which are slippery. The brakes must be set firmly either by holding the brakes normally or by having a reliable parking brake set. During the run up be continuously alert for any creeping forward which might occur! Check engine all pressures and temperatures!

Set the RPM and proceed with the runup in accordance with the checklist. Comply with the limits for mag drop. When a constant speed prop is involved -- the prop RPM should be cycled 2 or 3 times. This prop check is to assure that the prop governor is operating properly, to circulate oil thru the governing mechanism and to note that there is no oily mist thrown back over the engine cowling or the windshield. A fine oily mist on the windshield or cowling is the first indication that a prop seal may be leaking. In the event that the prop seal fails in flight you should expect a flood of dark oil over the windshield which will set you up for flying with visibility out the side windows only! As soon as the run up is completed reduce the RPM to 1000 or less. If any part of the runup is unsatisfactory, consult the checklist and go thru the complete runup a second time. In the event that the second runup is unsatisfactory, taxi back to the ramp and contact an A & P mechanic. Bottom line is -- if it doesn't check out on the ground it isn't safe to fly!

TAKE-OFF

Complete the **BEFORE TAKEOFF CHECKLIST.** When cleared for takeoff taxi so that you use all the reasonably available runway.

Operating to or from a strange airport and night operations anywhere will pose the risk of landing or taking off on the wrong runway. Airline crews have made this mistake. Military pilots have made this mistake as have general aviation pilots. If you are to avoid one of these dangerous traps you must use precautions including those mentioned here. When your landing light(s) illuminate a cement mixer, dump truck, or a big excavation dead ahead you are in deep trouble. You can see the wisdom of checking notams! Many years ago a wide body transport landed on the wrong runway at night at Mexico City. The runway was under repair or construction. The aircraft crashed into all sorts of equipment, went sliding down thru the area with great damage. This accident was attributed to a variety of factors on the ground and in the air. It was a nasty situation. Check the directional gyro against the number of the runway to be used for landing or takeoff. There are many past instances of aircraft landing or departing on a runway 15, for example, when the clearance was for runway 18 or maybe a runway 11. These errors are very dangerous! In a few instances aircraft have landed or departed down closed runways only to crash into barricades, holes, construction equipment, etc. Be sure you understand the runway in which you are to takeoff. If there is any doubt re-check with ground control or the tower. Accidents have happened at night, in daylight, on strange airports and at *"home"* fields. So -- always check the directional gyro before you start the takeoff roll or the approach. Maintain this habit. Additionally this check will detect an error in a magnetic compass caused by a headset too near the compass or some magnetic component amongst the baggage/cargo! Keep in mind that the runway alignment usually varies slightly from the runway number. Runway 27 may have an alignment of 274 degrees or 267 degrees.

A quick check of the actual runway alignment can be made by checking the airport diagram of a JEPP ILS approach chart where the actual magnetic course of each runway is displayed.

Slowly and smoothly align the plane on the runway centerline then hold the brakes while takeoff power is applied and a power check is made. When all power indications are observed to be normal, release the brakes and you are on your way. This whole sequence will require only a few seconds. On takeoff make a second power check in advance of rotate speed. Begin rotation by lifting the nose at the IAS which will cause the lift off to occur at the IAS stated for the subject aircraft. A review of several single engine POHs showed that each of these prescribed lift off prior to V_x IAS by widely varying amounts. It is then imperative that you study **your POH** takeoff and climb information in great detail in order to conform to the manufacturers recommendations. The POH must be complied with if the takeoff distance charts are to be valid and meaningful. If the proper lift off speed is used it is then reasonable that the aircraft will accelerate to or near V_x IAS as it climbs out of ground effect. Acceleration should normally continue to V_y as the climb continues. Rotate to the nose up pitch attitude which will provide climb at the best **rate** of climb (V_y) airspeed. This nose up pitch attitude can be established by visual reference of the aircraft nose vs. the natural horizon or by use of the pitch altitude indicator. The range of nose up pitch will usually vary from 8 to 10 degrees. I encourage the use of the pitch attitude instrument for maintaining the proper climb pitch attitude. If you elect to go to or very near V_x speed before beginning rotation -- in situations such as a takeoff in very rough air, a crosswind or during departure from a high altitude airport -- you should make an addition to the takeoff ground roll and climb to 50 feet which is set out in the POH charts.

I have found the addition of 15 to 20 percent to the POH numbers to be a reasonable adjustment for rotating at or near the V_x speed. If you are operating from runways where length far exceeds the minimum required amount -- then length may be no problem. However, be aware of the runway length requirement in every operating situation including intersection takeoffs.

A note of caution: If your tricycle gear aircraft is subject to *"WHEELBARROWING"* -- Main wheels lifting off the runway, at high speeds, while the nosewheel remains on the runway -- you should use slower takeoff and touchdown speeds to avoid this situation. Directional control is likely to be lost. Check this with a good CFI who is familiar with your make/model of aircraft.

POWER CHECKS DURING TAKEOFF

Normally Aspirated Engine - Fixed Pitch Prop

A good guide for a takeoff power check is to have RPM well within the cruise RPM range just before reaching rotate speed. The variables of prop pitch, engine condition, airport elevation, O.A.T. and mixture setting will all influence the takeoff RPM. If the RPM indications are not normal -- abort the takeoff. Repeat the run up and the entire checklist. If the abnormality persists have it remedied before attempting another takeoff. Make note of the normal RPM for your engine when approaching lift off speed. Consult the POH for cruise RPM range.

Normally Aspirated Engine - Constant Speed Prop

Apply takeoff power smoothly and positively. A power check consists of manifold pressure and RPM and if the engine is fuel injected -- a check of fuel flow. If any of the indications are other than normal, immediately abort the takeoff and repeat the run up and the entire check list.

If the engine continues to operate improperly, consult an A & P mechanic before attempting another takeoff.

TURBO-Charged Engines

Always use great care to apply takeoff power smoothly with any turbo charged engine. An excellent practice is to apply the power in two steps with a short hesitation between steps as follows: First gradually and smoothly advance the manifold pressure to 25 inches, hesitate momentarily until the turbo impeller has accelerated -- then advance the manifold pressure to slightly below the maximum takeoff manifold pressure setting. Thereafter make any final adjustment which may be needed. This procedure will prevent an overshoot of the manifold pressure during the time that the impeller speed is stabilizing. Immediately after the takeoff power is stabilized make a power check to determine that the engine is developing full power. This check includes manifold pressure, RPM and fuel flow. Maintain a familiarity with each of these indications so that abnormalities can be detected promptly. Smooth noise does not guarantee that the engine is producing takeoff power!

INITIAL CLIMB

During climb out it is important to monitor pitch attitude, IAS, rate of climb, engine performance and heading. This will be very helpful for night or low visibility operations and is essential for IFR operations. For aircraft equipped with retractable gear -- the gear should be raised only after positive rate of climb has been established and when there is insufficient runway remaining to land and stop on the runway. If flaps are extended they should be retracted while the desired pitch attitude and climb profile are maintained. See FIGURE 6-1 and FIGURE 6-2.

In all aircraft maintain best **rate** of climb speed (V_y) and full takeoff power until 1,000 feet above the departure airport. In departures from airports surrounded by high terrain or other obstructions it may be advisable to continue the **INITIAL CLIMB** to a higher altitude such as 1,200 feet or 1,500 feet or more before transitioning to **CRUISE CLIMB**. Do not *"baby"* the engine by an early power reduction! The engines are made to run if temperatures and pressures are in the green. The advantages of making the **INITIAL CLIMB** as outlined are numerous. Note the following:

1. Most power failures during the climb out occur at the time of the first power reduction!
2. Keeping your hand on the throttle(s) will avoid a pilot induced *"power failure"* at low altitude. Early power reductions have caused many power failures, at low altitude, because the pilot was operating the mixture control thinking he was operating the throttle or the prop control! Also, throttles have crept toward closed while being undetected.

During initial climb maintain appropriate drift correction in order to make good the desired ground track. If cleared to *"maintain runway heading"* -- do so!

COULD THESE ACCIDENTS HAVE BEEN PREVENTED?

A Piper PA-24-180 was on initial climb out when the engine lost power. In an attempt to turn back to the airport the aircraft stalled and crashed. The pilot was killed and the passenger was injured. Investigation revealed that the pilot executed the takeoff with the fuel selector positioned to the right tank which contained only residual fuel! Note: Always use the **before takeoff check list** which includes checking fuel selector position and quantity of fuel in the tank selected.

IF YOU SUFFER A PARTIAL OR TOTAL POWER FAILURE AFTER LIFT OFF AND THROUGH INITIAL CLIMB YOUR BEST BET IS TO LAND STRAIGHT AHEAD OR MAKE A SMALL TURN TO ALIGN WITH THE BEST AVAILABLE TOUCHDOWN AREA. ATTEMPTING TO GET BACK TO THE RUNWAY OR EVEN THE AIRPORT PROPERTY IS EXTREMELY DANGEROUS. TURN A DEAF EAR TO ANY AND ALL CLAIMS THAT THE TURN BACK MANEUVER CAN BE DONE SAFELY. THE FAA HAS A SAFETY BULLETIN ON THIS SUBJECT. THE TITLE IS *"THE IMPOSSIBLE TURN"*.

A Cessna 182 had power failure during takeoff. The airplane crashed on the airport property. The pilot and passenger were killed. A witness reported the engine was sputtering, during the takeoff roll, following a loud bang and surging. When airborne, the aircraft began a left turn then rolled into a right turn with a bank of approximately 90 degrees at which time the right wing tip and the nose hit the ground. This incident highlights the need for a power check on the takeoff. A takeoff abort would have avoided this accident.

An amphibian made a touch and go water landing. During initial climb the pilot reduced power and the aircraft descended back into the water. The aircraft struck hard, bounced into the air and hit the water a second time. The aircraft was substantially damaged and the pilot and three passengers were injured. NOTE: After takeoff leave maximum takeoff power applied until 1,000 feet above the ground or water. Additionally know and maintain the appropriate nose up pitch attitude for climb at best **rate** of climb speed.

A Beech A-36 pilot was taking off when the cabin door came open. At Approximately 300 feet AGL the aircraft made a turn left followed quickly by a turn right. The nose pitched down and the aircraft crashed and burned.

The pilot, who was the only occupant, was killed. The NTSB determined that the probable cause was the pilot's failure to maintain control of the aircraft because he had diverted his attention to dealing with the open door. The pilot was high-time in the A-36.

Attempting to close an open door in flight is a very dangerous activity and virtually impossible. Some pilots claim that the doors can be closed in flight. Others have heard stories about a *"sure way"* to accomplish this feat. Don't believe it and don't try it! Do not be suckered into any of this type of thinking because it is very unsafe. If a door comes open it usually happens between the start of takeoff roll and 1,000 feet AGL. The door opening in flight begins with the sound of air rushing, then a bang as the door comes open. Thereafter the trailing edge of the door is held open about 3 inches by the airflow. The resulting wind noise will be very loud. The circulation of air in the cabin will be very strong. Loose papers, charts, etc. may be blown out of the cabin. Remember **FLY THE AIRPLANE** and return to the airport. Check the POH for any notes or limitations which apply when operating with the door open. Plan your approach so that it is on profile and appropriate speed. The incident should terminate with a normal approach and landing. Now close and secure the door once the airplane is safely on the ground.

Another incident occurred where a cabin door came open shortly after takeoff. The pilot notified the tower and was cleared to land on any runway. While the aircraft was maneuvering to return for landing, it abruptly flew into the ground. The pilot and two passengers were killed while a third passenger was seriously injured. This aircraft was a Beech 24-R SIERRA.

The POH had a procedure for door unlatched in flight. Because this procedure is typical it is repeated here and it states -- *"If a cabin door latch is not fully engaged it may come unlatched in flight. This may occur during or just after takeoff. The door will trail in a position approximately 3 inches open. A buffet may be encountered with the door open in flight. Return to the field in a normal manner. If practicable, during the landing flare-out, have a passenger hold the door to prevent it from swinging open."* The *pilot* should close, latch and check cabin door(s) utilizing extra care and caution.

A Cessna R172K failed to develop full takeoff power during the takeoff roll. The pilot did not abort. The aircraft struck trees approximately 500 feet beyond the runway and crashed. The aircraft was substantially damaged. The pilot and passenger received minor injuries. Investigation revealed that the mixture control was improperly adjusted resulting in lessened power output. NOTE: A power check at the beginning of the takeoff roll would have alerted the pilot and prevented this accident. Engine noise and the feel of acceleration are not accurate and reliable indicators of normal power output. A power check must by made by the appropriate engine instruments with the assurance that the indications are truly normal for takeoff power!

THESE TWO GOTCHAS - GOT ME

A few years ago I was taxiing out for departure in a Piper PA-28-300 Lance when I noticed that the magnetic compass was off about 20 degrees. What was wrong here? The culprit was a portable public address system which I had loaded into the nose baggage pit. Magnetized components of the PA system were pulling the magnetic compass to an erroneous indication.

About 1983, my wife and I were departing Denver-Centennial airport for Great Bend, Kansas to spend the day with relatives. The time was around 10:00 a.m. on a winter day with the O.A.T. at 40 degrees F and severe clear. We were flying a Bonanza S-35. During the before engine start I had some difficulty getting the door closed and latched. I made a second try at closing the door and still had difficulty but decided that since I was not familiar with this airplane -- the door had to be OK. Wrong! We launched down the runway, lifted off and as I reached for the gear handle there was a shrill whistle followed by a loud **BOOM** and blast of cold air as the door came open to about 3 inches at the trailing edge. The forward cockpit was enveloped in a swirl of dust, bits of paper, cigarette ashes, etc. The rush of cold air was numbing to my wife who was wearing a light jacket. I was not going to try to close the door in flight. I continued flying the plane. After reaching about 1,000 feet above the airport I slowed down to 100 kts. called the tower and was cleared to land on any runway. I chose 17L because it was longest and widest. We landed without incident. We waited on the ground long enough to get warm again and really close and latch the door. We continued the flight without further difficulty. My impatience had caused an unnecessary problem, great discomfort for my wife and excitement I didn't need!

SUMMARY

The runup, takeoff and initial climb are the operating segments which spawn serious and often fatal accidents. A professional attitude, disciplined use of checklists and good operating practices will prevent virtually all accidents during these areas of operation.

Takeoff Profile for Single Engine Aircraft

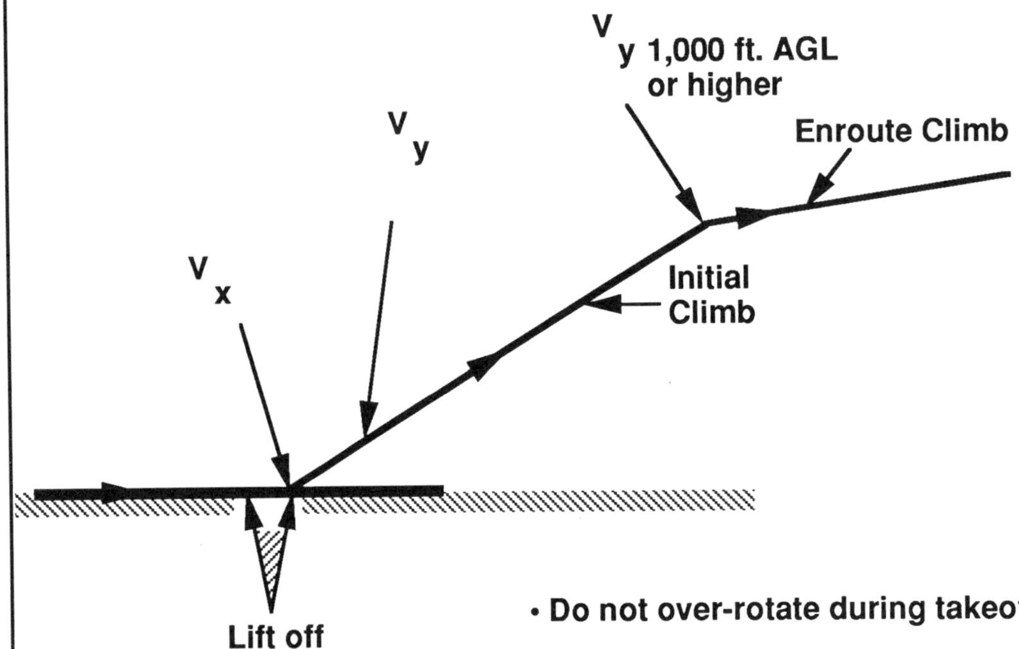

- Do not over-rotate during takeoff!
- Refer to POH for Vx and Vy speeds.

- Begin rotation at the appropriate IAS.
- Rotate to the pitch attitude for climb at Vy (Note this pitch attitude for the airplane you fly).
- Vy will usually be attained at about 75 ft. AGL.
- Note the normal rate of climb during each initial climb. If the climb rate is significantly below normal, check the power setting, gear, flaps, etc.

Figure 6-1

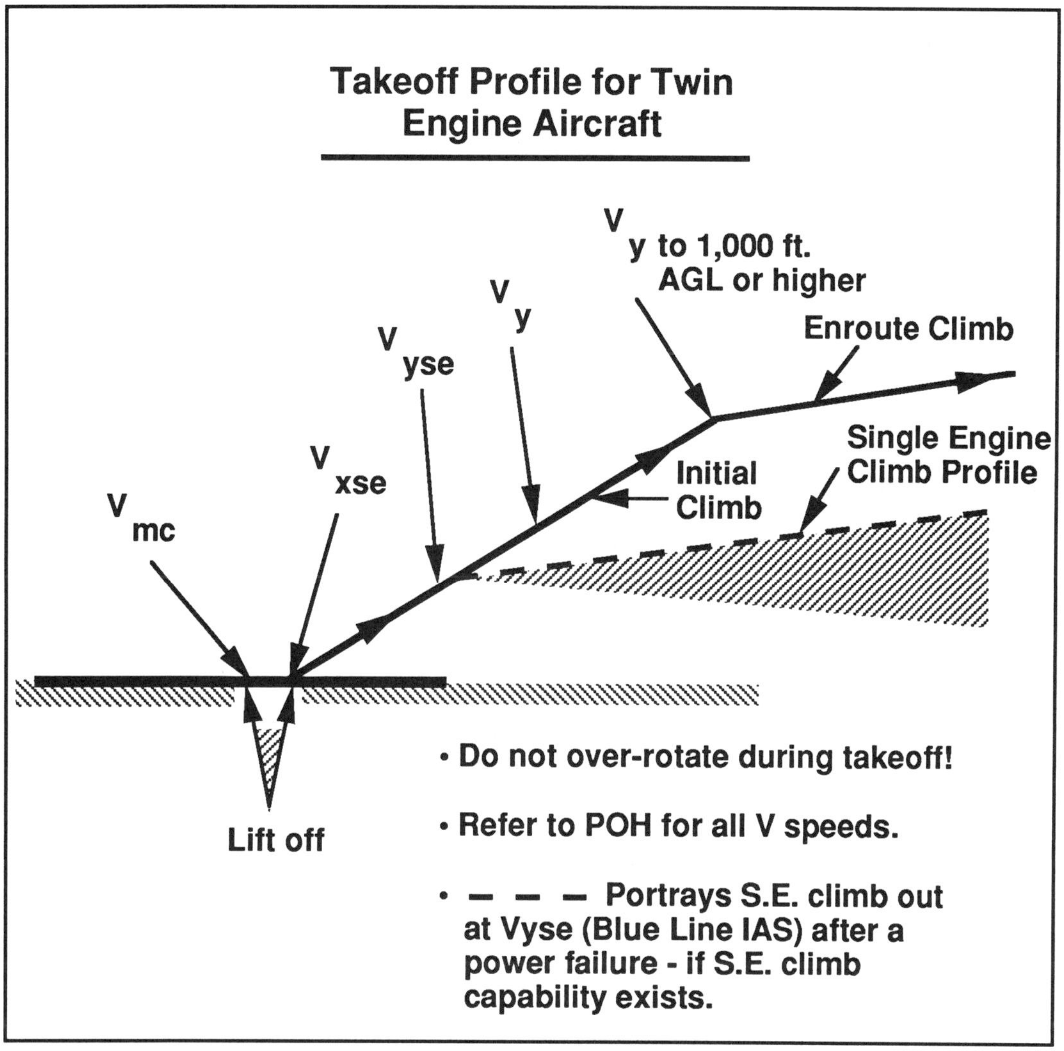

- Begin rotation so that lift off will occur at Vxse.
- Rotate to the pitch attitude for climb at Vy (Note this pitch attitude for the twin you fly).
- Vyse will usually be attained at about 50 ft. AGL.
- Note the normal rate of climb during each initial climb. If the climb rate is significantly below normal, check power settings, gear, flaps, etc.

Figure 6-2

This chapter begins when the aircraft transitions from initial climb to enroute climb and ends when the aircraft is in the terminal area ready to begin the approach and landing.

ENROUTE CLIMB

At 1,000 feet above airport elevation or higher establish enroute climb by lowering the pitch attitude enough for the airspeed to increase to the desired climb speed. At that point set climb power and accomplish the AFTER TAKEOFF check list. Continue the climb while leaning the mixture as appropriate -- refer to the POH. Upon reaching cruise altitude leave climb power set until leveled out, cowl flaps are closed and the airspeed has increased to cruise IAS. At this point set cruise power and adjust mixture. Trim as needed. Accomplish the CRUISE CHECKLIST. Now fill out the flight log i.e. time off, ETA for the next check point and start the fuel burn out log. Flight management will become increasingly important as the flight continues.

CRUISE

The cruise phase of flight is where the overall operation must be monitored while navigating and communicating. Careful attention to enroute weather, ground speed, fuel burn out and the flight log are essential. Do not just *"lean back and let it happen"*! If you are not busy during cruise flight then something is being neglected. Experience tells us that the cruise phase of flight is where serious problems frequently develop. The auto pilot may be inadvertently disconnected, alternators may fail, vacuum pumps die, engine malfunctions occur -- the list goes on and on. For night and IFR operations equipment failures impact the operation more severely. Be alert -- be a manager!

An additional problem seldom addressed is the medical emergency which may arise while enroute.

When confronted with a medical emergency you should contact the nearest communication facility and describe the problem. State what airport you are diverting to, the ETA and request ambulance and medical personnel to meet the aircraft. Be sure to state which FBO you are going to so that the ambulance can meet you without delay. Descend to a lower altitude. The application of oxygen may be indicated for the stricken passenger. ATC, approach control, tower and UNICOM are resources of assistance. Ask them to help. Request priority handling or declare an emergency as seems appropriate. Think CRM and utilize it.

DESCENT PLANS

For a descent **in smooth air** -- the descent should be made at 500 FPM with cruise power adjusted throughout so that the airspeed remains in the safe range. In all descents -- nose down trim must be adjusted several times if descent speed is higher than cruise speed. Once the airspeed has *"topped out"* and trim is set, get the ATIS, check the fuel quantity, set the tank selector as appropriate and adjust the mixture. Accomplish the DESCENT CHECKLIST.

A second method for descent is to reduce power so that the descent is made a 500 FPM **and at cruise airspeed**. This method virtually eliminates the trim activity and would be appropriate in very light chop or as a pilot preference. Accomplish the DESCENT CHECKLIST.

The third method **is for rough air**. Again, the descent rate should be 500 FPM at approximately 90% of maneuvering speed for the aircraft weight. If a retractable gear airplane is being operated a good option, to control airspeed during the descent, is to slow down and extend the landing gear before beginning the descent. Accomplish the DESCENT CHECKLIST.

The pilot then has only to adjust power and trim to establish the desired airspeed and the 500 FPM rate of descent. This procedure is frequently used in the mountains for descent from a pass into a valley when the air is very rough. Typically in a Cessna T-210 or Bonanza this type of descent would be made with gear down and IAS around 110 KT. At these lower airspeeds the airplane can tolerate the heavy bumps without damage and the occupants will be more comfortable. Check the POH for your aircraft to determine the maneuvering speed for the gross weight at which you are operating. My preference is to operate at a speed approximately 10% below the applicable maneuvering speed in order to assure an adequate safety margin.

WHEN TO BEGIN THE DESCENT

When and where to begin the descent from cruise altitude? First -- use a descent rate of 500 FPM as the standard for all descents from cruise altitude. The formula then becomes: Ground speed (miles per minute) X 2 = miles traveled for each 1,000 feet of descent X thousands of feet to descend = distance during descent.

The following is a typical example:
The ground speed during descent is estimated to be 150 Kts. and the altitude between cruise and the destination traffic pattern is 6,000 ft.
The computation is:
150 Kts. equals 2.5 NM per minute X 2 equals 5.0 NM per 1,000 ft. of descent X 6 for 6,000 ft. descent, equals 30 NM as the distance required for the descent from cruise altitude to the traffic pattern altitude. You will need some additional distance to SLOW DOWN AND MANEUVER for the traffic pattern. I usually add 3 to 5 NM for this part of the operation. Thus, in this situation I would begin my descent 33 NM from the destination airport.

It is obvious that a variety of adjustments will be required to accommodate the many VFR and IFR arrival situations which may be encountered. Skill in using this descent plan will increase as experience is gained. The most prominent error during descent operations is to arrive in the traffic pattern area or the IFR approach area too high and too fast. Allow reasonable distance to slow down and maneuver at the bottom of the descent from cruise.

If, when under ATC control, you are approaching your point of descent and have not received a clearance to descend call ATC and say *"I would like lower"* or *"I request descent to ____feet"*. These communications will assure that the descent can be started at the desired point. If the descent from cruise altitude is begun later than desired, the pilot will need to increase the rate of descent to more than 500 FPM as is necessary for the situation. Power will then need to be reduced to a much lower setting which sets the stage for *"shock cooling"* the engine and possible passenger discomfort. The least desirable situation is to remain at cruise altitude far past the desired descent point then be cleared -- *"dumped"*, to a lower altitude. These situations involve a need for unreasonable rates of descent which may be impractical for your aircraft. If you get caught in such a clearance reduce airspeed during the descent and adjust to the situation as best you can. Advise ATC of your situation.

COULD THESE ACCIDENTS HAVE BEEN PREVENTED?

A Cessna 182 RG was on a day VFR flight from California to Montana. During cruise at 6,000 feet MSL, the engine lost power. The aircraft was destroyed during an emergency landing on rough terrain. The pilot and passenger were not injured.

Examination revealed that the right fuel tank was empty while the left tank contained 23 gallons of fuel. The fuel selector was found positioned to the empty tank. This accident may have been avoided by keeping a fuel burn out log current or by utilizing the engine failure or power loss procedure in the POH.

A BEECH A-36 was on a cross-country flight when the engine quit in the vicinity of the destination. The aircraft crashed in a field short of the airport and the aircraft was substantially damaged. The passenger received minor injuries. The pilot was unhurt. No fuel was found in the tanks. Where was the fuel burn out calculation? Where was the reserve fuel that is required by regulations and common sense? This accident could have been prevented by good flight management and an enroute fuel stop.

A private pilot and his passenger were flying a Cessna 206 to Alaska for a hunting trip. During the final leg of the flight the aircraft struck a mountain glacier at the 5,600 foot level. The occupants were killed and the aircraft destroyed. Weather conditions at the time were estimated to have been lower than VFR with rain and clouds obscuring mountains and severe turbulence. The NTSB report noted that the pilots former flight instructor had reportedly advised him not to undertake the flight due to his lack of experience and the potentially severe weather conditions in Alaska. The pilot had logged a total of 164 hours and was instrument rated. This may have been an example of proceeding VFR in IFR conditions. Minimal flying experience was a likely contributing factor. My recommended experience level for this flight would have been 1,000 total pilot hours including 500 hours of cross country flying.

During a night cross-country flight from a Wyoming airport, a Mooney M-20 crashed into a ridge at the 9,000 foot level. Instrument conditions prevailed at the time and no flight plan was filed. The pilot was killed and the passenger was seriously injured. The passenger stated in part that they were on top of the clouds when they ran into snow. He then observed ice on the wings and said they were in the clouds for quite a while. The passenger went on to say that the engine started missing and finally died. This appears to be an example of departure via special VFR clearance out of the local area and on into IFR conditions with no flight plan or clearance. The aircraft was not equipped for operation in icing conditions. This flight should have been cancelled regardless of the consequences of the delay. Had the pilot utilized Cockpit Resource Management during pre-flight and flight planning he would have been led to the decision to stay on the ground.

A Cessna 150 struck trees and terrain west of the summit of a pass in Washington. The pilot and his wife were on a trip to Alaska. One witness reported seeing the aircraft fly into a thick layer of low clouds moving up from the valley below. The pilot and his wife sustained fatal injuries in the crash. No record was found of the pilot having obtained a weather briefing and no flight plan had been filed. This may be another case of inadequate pre-flight planning and the pilots decision to *"try to get through"*! There is the saying -- *"Nothing I need to do tomorrow is worth risking my life today"*!

SUMMARY

Additional enroute difficulties involve engine problems and infrequently electrical system or avionics failures. The engine failures involve mechanical failures such as connecting rod or valve breakage. These and other failures are caused by loss of oil when oil lines or fittings break. Vibration and age take their toll of all engine lines, fittings, and hoses. Vacuum pumps should be carefully monitored for hours in service. It is my experience that the operating life of the average vacuum pump is 500 to 700 hours. Don't run a pump until it fails! If you fly IFR or night cross country a standby vacuum system is a necessity for safe operation.

Enroute weather related accidents continue to be a major killer of our fellow pilots -- why? The answer lies in the fact that the cross country training requirements for a Private Pilot License are woefully inadequate. To be proficient enough to fly more than 100 miles from the home airport a pilot should have at least 12 hours of cross country **dual** training. This training should consist of at least one triangular route with 4 hours flying time on each leg. There should then be landings at three airports on each leg. Yes, this will cost some money and require an overnight stay. However, this would be an invaluable real life experience. As an alternate plan hire a good CFI to accompany you on the first business or pleasure cross country trips. The tragedy of acquiring a Private Pilot Certificate is that the new pilot, who was a Student Pilot yesterday and restricted from carrying passengers, is today a Private Pilot who may legally load up his family or business associates and depart on a trip to any point in the U.S.. This individual has neither the training or experience to operate such a flight.

Weather is very different from one geographical area to another. The weather may be clear and 60 miles in the Rocky Mountains, while entirely different weather prevails in the Pacific Northwest, the Upper Great Plains, the Gulf Coast, Desert Southwest, Great Lakes area, the Ohio and Mississippi River Valleys, the Eastern U.S. etc.. There are about a dozen unique weather areas in the U.S. Additionally the terrain clearance and navigational problems vary widely from area to area. Cross country flying, if it is to be safe, requires training and experience gained while operating in a variety of conditions. In the present system the new or low time pilot is *"thrown to the wolves"* on the subject of safe and orderly cross country flying. Be sure to augment your Private Pilot Certificate with additional cross country training. Thoughtfully examine your capabilities and get additional training if needed. Ask yourself if you are as competent as you desire or have you been taking chances with the weather and *"lucking out"*. Let's work together to avoid weather situations which present any element of operational risk!

In this chapter *"Approach"* will encompass airplane operations beginning with maneuvering in the vicinity of the airport thru the landing approach, the landing and the roll out.

PRE-APPROACH MANEUVERING

As you approach a traffic pattern the items for attention are situational awareness, good communications, use of checklists, careful scanning and speed control. The main culprit in this list is speed control especially for pilots operating high performance aircraft. Avoid being too fast when close in to the traffic pattern. Anticipate the desired speed and have in mind a power setting to use for that speed. For example when about 3 miles from the downwind leg, in a Bonanza, Cessna 210, Mooney, Piper Arrow and similar aircraft, set the manifold pressure at 16-17 inches. Now maintain pattern altitude while the aircraft slows down. Note that this is close to the manifold pressure required to fly straight and level gear and flaps up at 110-120 Kts. IAS.

THE APPROACH

When opposite the touchdown point in the traffic pattern, Figure 8-1 (Step 1) or approaching the 3 mile point during a straight approach, Figure 8-2 (Step 1) extend gear and flaps as desired. Let the aircraft slow to approach speed with a descent rate of about 500 FPM. Make adjustments to power as needed.

Figure 8-3 is to be used by you to make a table of power settings for the aircraft you fly. Fly your plane thru a stabilized visual approach and fill in the numbers. Use these numbers to develop your standardized approach profile.

After a few practice stabilized visual approaches you will find that you have an orderly approach with airspeed and descent nicely controlled. You will in fact be doing a professional airline type approach to a touchdown at around 1,000 feet beyond the runway threshold. The descent will duplicate a 3 degree ILS glide slope descent profile or a 3 degree VASI descent profile. All final approaches will be standardized and identical whether day, night or during the visual approach portion of any IFR approach! Practice this on your own until you have the technique and the numbers that work well.

See Figure 8-4 for sample tables of approaches -- one for a Cessna 172 and one for Beech Bonanzas. These will be approximate for similar aircraft i.e. fixed gear with flaps and fixed pitch prop or retractable gear with flaps and constant speed prop. Keep in mind that each approach consists of a series of adjustments to airspeed, rate of descent and power. It is obvious then that a well planned approach will require a minimum of adjustments and thus it will be more precise and easier to accomplish. A major advantage to the precision visual approach is that it establishes the landing touchdown point where the pilot desires it to be -- not a hit or miss situation. Passengers will feel confident and safe during a precision approach because of the orderly execution of the approach.

APPROACH-WORDS OF CAUTION

If you are confronted with excessive cross wind, heavy turbulence, thunderstorms within a few miles of the airport or slippery runways discontinue the approach. The options then to be considered, include use of a different runway, go to a suitable nearby alternate airport or return to the departure airport.

Anytime your destination airport is a one runway field it is wise to carry an extra fuel reserve so that you can hold for 30 or 40 minutes in the area or go to an alternate airport and land with more than a minimum fuel reserve.

For example, I was turning on to final approach at a one runway airport in Wisconsin in the early 1950's when a Navion landed gear up just ahead of me. At a small airport you should expect a delay of at least one hour in a situation such as this. I diverted to the nearest suitable airport some 25 miles away. I was flying an old Lockheed 10-A on a charter trip. This diversion took nearly 20 minutes and a good deal of my extra reserve fuel supply.

In the mid 1980's I was turning on to final approach at Jackpot, Nevada when the plane landing ahead of me suffered the collapse of the nose gear during roll out. I announced, via UNICOM, that we were going to exit the traffic pattern and hold awaiting the outcome of removal efforts. A quick computation of fuel burned and fuel remaining indicated that we could hold for 30 minutes and thereafter go to Twin Falls, Idaho, 35 miles north, and land with a comfortable fuel reserve. In this instance there were a good number of people on the ground who promptly moved the damaged aircraft from the runway. We were able to land after a hold of some 20 minutes.

In a third instance, my friend Bob Akers and I arrived at Douglas, Wyoming on a winter morning after a 12 inch snowfall. The single runway was to have been plowed. However the snowplow engine was being difficult to start so at our arrival the runway was still closed.

The owner of the Cheyenne II turbo-prop, our friend Carl Williams of Denver, awaited patiently in the FBO office while we flew a VFR race track holding pattern in the valley. We kept abreast of developments via UNICOM. The nearest useable alternate was Cheyenne, Wyoming. Having plenty of fuel we decided to stay in the area for awhile and as it worked out, the snow plow was started and a strip was plowed the length of the runway and into the ramp. We were on the ground after a delay of some 45 minutes.

Keep in mind that many surprises encountered in flying can be handled safely -- **if** the pilot has some extra fuel! The FAA minimum fuel reserves for day, night and IFR operations are just that, **minimums.** In the real world these FAA minimums are woefully inadequate. The responsible pilot will have on board, at all times, a fuel reserve, to accommodate contingencies which may arise. Running out of fuel is inexcusable!

THE LANDING

A good approach is the foundation for a good landing. The following discussion applies to normal landings and not to short field landings which may require some adjustments in technique, configuration of the aircraft, approach speeds, etc. The approach planning should be in accordance with the guidelines set out for the PRECISION VISUAL APPROACH illustrated in FIGURE 8-1 and FIGURE 8-2.

THE AIM POINT AND TOUCHDOWN POINT

The aim point should be about 500 feet beyond the runway threshold or the displaced threshold. The touchdown point for a normal landing should be at about 1,000 feet beyond the runway threshold or the displaced threshold. See FIGURE 8-5.

After touchdown the landing roll at lower elevations airports will range from about 1000 feet to 1,400 feet for single engine aircraft, Cessna 152's thru Beech Bonanzas. Check the POH for the aircraft you fly. Keep in mind that you will normally use less braking than the test pilot who generated the numbers for the POH. To compensate for the average normal braking I suggest that you add about 25% to the POH ground roll numbers. Having a planned touchdown point of 1,000 feet plus 1,400 feet for ground roll, the **minimum** runway length requirement then becomes 2,400 feet at low elevations. Check your POH and if you have difficulty interpreting the numbers in the POH or adjusting them to your normal operations -- check with a CFI familiar with your aircraft.

THE LOOK POINT- What Is It?

The LOOK POINT refers to where the pilot concentrates his/her vision during the landing flare, the landing and the roll out. Visualize a cone of vision which is projected from the pilot's eyes to a point ahead of the aircraft. The angle of the cone remains constant throughout the landing sequence -- thus the area of concentrated vision moves steadily forward with the aircraft. This LOOK POINT is similar to the illuminated area provided by a landing light during a night landing. See FIGURE 8-6.

WHAT CAUSES BAD LANDINGS?

Bad landings frequently result from bad approaches. More disturbing however, is the fact that where pilots look during landings appears to be a hit and miss situation. When I was learning to land, my recollection is that my flight instructor said only *"look a little ahead and slightly to the left."* Somehow I learned through trial and error.

When I began to get it right I apparently developed a sense of what was about correct and had fair luck at making good landings most of the time. Occasionally I would be completely surprised by dropping it in unexpectedly. Conversely, I would sometimes be confident that I was about to accomplish a *"greaser"* only to hit the runway hard, bounce all too high and need to do a bad landing recovery. What was going on here? In 1942 I was being checked out in an open cockpit WACO UPF-7. On the first practice landing I leveled off very high! The plane was in near 3-point attitude and about 3 feet off the ground -- while I thought we were about to touchdown! The instructor called for a go-around and when underway he remarked -- *"I bet you were looking at the far end of the field -- if you look too far ahead you will drop it hard every time."* Ah ha, now the mystery is beginning to unfold! I thought back at the times I had *"dropped"* Aeronica Chiefs, Porterfields, Rearwins and J-3 Cubs. Now I know that the cause was my looking too far ahead of the aircraft. But how far is too far? Was I looking at the ground too close to the aircraft when I had previously flown into the ground unexpectedly? My curiosity was leading me to be serious about developing a LOOK POINT that would be applicable to all aircraft. I was approaching my CAA (circa 1942) Flight Instructor check ride and I wanted to do well and thereafter have a good teaching tool so I continued to use and refine the LOOK POINT plan.

HOW TO LAND USING THE LOOK POINT

As the aircraft crosses the runway threshold or the displaced threshold the wheels should be approximately 12 feet to 15 feet above the runway. The aircraft should be configured for landing with appropriate airspeed and rate of descent. If there are obstructions adjust accordingly.

After the aircraft crosses the runway threshold and during the landing flare the use of the LOOK POINT begins. The LOOK POINT should be ahead of the airplane -- and slightly left of the centerline axis of the fuselage so that the engine cowling will not block out the LOOK POINT view as the landing progresses.

The LOOK POINT for aircraft such as Cessna 152, 172, 182, 210, Bonanza, Mooney, Bellanca Viking etc. should be approximately 1,000 feet to 1,200 feet ahead of the aircraft. These aircraft have pilots' eye heights above the runway within a range of about 5 feet. Keep in mind that these are not finite numbers but rather general guidelines that are proven to work well, for me, over a period of more than five decades of operating experience. As a flight instructor in general aviation aircraft of all sizes from J-3 Cubs, Cessna 150-152 thru Stearmans, Cessna T-210, Bonanza, Cessna 414 etc., my trainees consistently make good approaches and good landings near the 1,000 foot touchdown point. The reasons are that they adhere to the PRECISION VISUAL APPROACH and utilize the LOOK POINT during the landing.

What then is the LOOK POINT for aircraft where the pilot's eye height is higher -- say in the range of 6 feet to about 8 feet? This range of aircraft would include the Staggerwing Beech, DeHaviland Beaver, Cessna, 195, 180, 185, Gulliwing Stinson, Stearmans (front cockpit) and most larger twins. The LOOK POINT range for this group and similar aircraft is then about 1,200 feet to 1,400 feet. It then follows that there appears to be a small range of angles from the landing surface to the pilot's eye wherein the pilot can very accurately judge the height of the wheels or the floats above the landing surface.

WHEN USING THE PROPER LOOK POINT THE PILOT WILL BE ABLE TO MAINTAIN PRECISE DIRECTIONAL CONTROL, BE AWARE OF THE NOSE UP PITCH ATTITUDE AND BE ACCURATELY AWARE OF WHEEL HEIGHT ABOVE THE LANDING SURFACE. THE THREE JUST MENTIONED ITEMS ARE THE THREE INGREDIENTS NECESSARY FOR A GREAT LANDING.

THE TOUCHDOWN AND THE ROLL OUT

During the moments just before touchdown the wheels should be within 4 to 6 inches of the runway. You should be in the process of *"holding it off"* until the aircraft is in a tail low/nose up pitch altitude as it touches the runway. After touchdown stabilize the directional control and keep your hand on the throttle(s) until clear of the runway. When clear of the runway and at a low speed it is then safe to retract flaps, open cowl flaps, etc. as is appropriate for the aircraft being operated.

DIRECTIONAL CONTROL -- DON'T LOSE IT

Each year there are dozens of accidents where directional control is lost during the roll out because of pilot attention being diverted to retracting wing flaps, opening cowl flaps, turning off the transponder or other ill timed actions. If a fellow pilot or a flight instructor has said to *"get the flaps up immediately"* after landing -- they have given bad advice. How much extra lift is being generated after the wing and flaps combination is well below stall speed? Answer -- virtually NONE! Then there is the argument that you must get the weight on the main gear so the brakes work better. First, if the wing isn't lifting the airplane the weight IS on the wheels! If the wing is 5 Kts below stall speed, on roll out, the difference in the lift generated by the wing with flaps extended versus flaps up will not be of any practical value.

However the extra aerodynamic drag of flaps extended will be beneficial immediately following touchdown. Lastly, if the approach and landing have been accomplished in accordance with the previously recommended methods there will be minimal need for the use of brakes during the roll out. Keep in mind that directional control during the landing and the roll out is of the utmost importance -- give it your undivided attention! Keeping your hand on the throttle(s) during this period will enable you to instantly apply power to increase rudder effectivity if needed. Additionally a firm grip on the throttle(s) will prevent some distraction like reaching for a wrong switch -- i.e. hand on gear switch thinking it is the flap switch. Don't rely on the SQUAT switch working!

SLIPPERY RUNWAY CONDITIONS AND BRAKING TECHNIQUE
When on a runway with patches of ice and snow it is important to avoid skidding a tire from a slippery surface to a bare area of the runway or taxi strip. If this happens you should expect to grind a flat spot on the tire. When operating on any slippery surface the taxi speed should be slow and brake applications should be well planned and very smooth. Should it be necessary to land on a runway covered with ice -- stay off the brakes and let the aircraft slow down gradually to a low speed, then apply brakes gently. Keep in mind that light braking is of some value only as long as the wheels are turning! If enough brake pressure is applied so that the wheels are locked up and skidding you then have a sled, with wings, which is out of control! The only recovery is to release the brakes and add some power -- about 1500 RPM -- to increase rudder effectivity long enough to regain directional control. A demanding situation is to land on an ice covered runway in a crosswind. It goes without saying that the crosswind must be somewhat light for this to be successful.

Make the touchdown using the wing down crosswind technique. At touchdown hold the aileron into the wind and increase the aileron displacement as the plane slows down. The underlying purpose here is to utilize the extra drag of the down aileron on the downwind wing, to assist in preventing the plane from weathercocking into the wind and skidding sideways. After the aircraft is slowed and clear of the runway proceed with caution. It may be advisable to shut down and be towed to the ramp or hangar by a tug. An airplane being blown across an ice covered ramp is a very helpless situation and potentially dangerous.

SUMMARY

Pick a good day and practice the PRECISION VISUAL APPROACH using the distance guidelines I have recommended -- come up with a LOOK POINT that works well during **your** landings. The net result will be smooth approaches which consistently bring you to touchdown at 1,000 feet to 1,200 feet from the normal or displaced threshold. The landing will be smooth and the roll out straight as an arrow. You can now forget everything you have been told about runway slope, look here -- look there, etc. All that rhetoric harkens back to the stone age of flying. Some of that material may work at the home airport and its individual conditions. However, for day and night operations with the endless variety of airports and conditions that may be encountered -- these old wives tales are worthless at best and downright dangerous at worst! In the chapters that follow on night operations, IFR operations and tailwheel operations the advantages of the PRECISION VISUAL APPROACH and landing plan will become increasingly obvious. Of all the segments of a routine flight -- I can think of none more gratifying than a well executed approach and landing. I call this the *"ART OF THE APPROACH AND LANDING."*

COULD THESE ACCIDENTS HAVE BEEN PREVENTED?

A Cessna 172 touched down long on the runway and came to rest in a creek. The aircraft sustained substantial damage and the pilot suffered minor injuries. The pilot stated that touchdown was about mid-field approximately 10 knots faster than the normal 65 knots approach speed for the Cessna 172. The pilot applied full brakes after touchdown but was unable to prevent the plane from running off the end of the runway. NOTE: Proper airspeed for approaches is 1.3 times the stall speed for the aircraft weight and flap configuration during the approach. Too much speed is to be avoided. Care should be taken to make timely and appropriate adjustments so that the altitude and airspeed are precisely controlled throughout each landing approach. A go-around would have saved the day in this situation.

A Mooney 20C struck trees and crashed while attempting a night landing. The passenger stated that during final approach white lights were observed. The airport is equipped with a 2-bar VASI system. The sighting of white lights indicates a high approach. The aircraft wreckage was located about 1500 feet beyond the **departure** end of the runway. This accident could have been avoided by altitude planning throughout the approach and by precise control of airspeed. When a final approach is too high it can quickly develop into a high and too fast approach as the pilot increases the rate of descent. These situations call for a prompt go-around.

A Piper PA-34-200T was on an IFR flight plan. Prior to takeoff the pilot received a weather briefing which called for widely scattered thunderstorms. On arrival at the destination a thunderstorm was over the airport.

An airline crew on an ILS approach ahead of the PA-34, made a missed approach due to windshear. The PA-34 pilot heard the tower report of windshear. The approach was continued and a strong downdraft was encountered which resulted in the aircraft being slammed to the runway. The aircraft was substantially damaged. The pilot and passengers escaped injury. **Thunderstorms, in the vicinity of the destination airport, must be avoided.** This situation called for a missed approach and request for a holding pattern for 15 to 30 minutes. This would have allowed time for the thunderstorm to move past the airport. The approach could then have been resumed with safety.

In Illinois a low time pilot was making a landing approach in a single engine Cessna 172XP when a *"rapid descent took place."* The pilot stated that he added power and lowered the nose. The nose gear and propeller struck the runway. The airplane bounced and received substantial damage. There were no personal injuries. A stabilized and carefully monitored final approach could have prevented this accident.

A pilot flying a Piper PA-28-180 was carrying passengers on a local sightseeing flight in northern Michigan and while flaring to land the aircraft ballooned due to excessive airspeed. The aircraft bounced twice and began to veer to the left at which point the pilot began a go-around. During the initial part of the go-around the left wing struck a runway edge light. The aircraft remained airborne and the pilot later made a safe landing. The wind was calm. The aircraft received substantial damage but the pilot and passengers were not injured.

This is another situation which could have been prevented by a stabilized approach which would have emphasized the airspeed at 1.3 times the stall speed for the weight and configuration of the aircraft during the approach. Additionally, directional control was not maintained during this go-around.

A turboprop commuter flight was on landing approach when it landed short of the runway threshold. The left main gear collapsed, a propeller struck the runway and the aircraft departed the left side of the runway into a snow bank. The crew and passengers were uninjured. The aircraft received substantial damage. The NTSB determined the probable causes to have been pilot misjudgment of altitude, distance, and airspeed. This is another accident which could have been prevented by a PRECISION VISUAL APPROACH!

A Beech Bonanza pilot had clearance for an ILS approach. At about 400 feet AGL he reported runway in sight but as he continued he lost sight of the runway due to fog. He was advised that visibility had dropped to 1/16 of a mile. He initiated a missed approach and a few seconds later the aircraft hit the runway. The aircraft was destroyed. The pilot and passenger received minor injuries. The pilot had high total hours and was reported to be highly experienced in Bonanzas.

When go-arounds or missed approaches are attempted many end up by descending onto the runway! Study the figure titled GO-AROUND at the end of Chapter 9. Note that the first two steps are accomplished simultaneously -- i.e. POWER AND PITCH.

Precision Visual Approach
Traffic Pattern

This rate of descent will approximate the descent path of a 3 degree ILS glide slope.

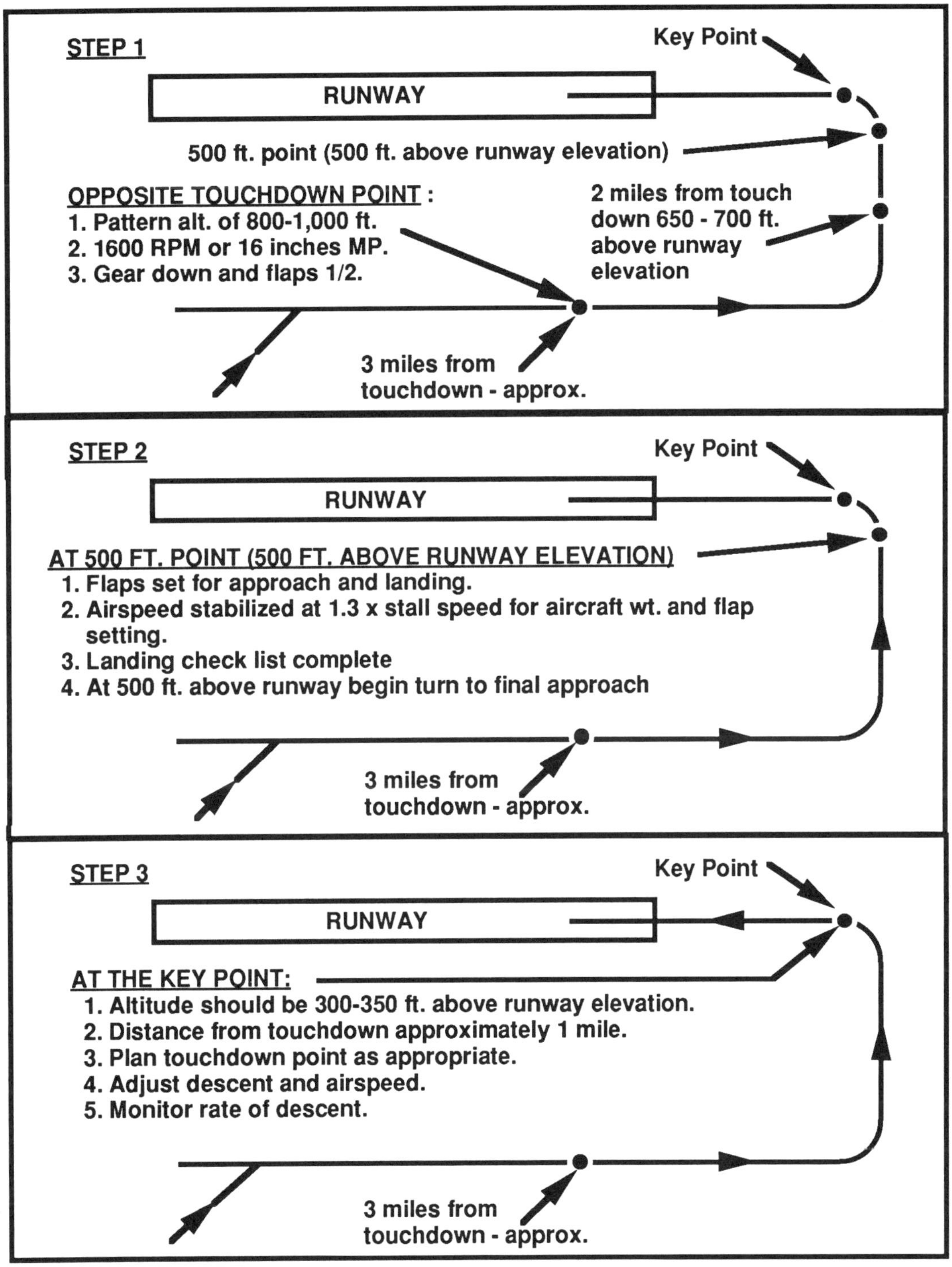

Figure 8-1

Precision Visual Approach
(Straight in Approach - VFR)

This rate of descent will approximate the descent path of a 3 degree ILS glide slope. Note that his approach is executed the same way as the traffic pattern approach in Figure 8-1.

<u>STEP 1</u> - AT 1,000 FT. ABOVE RUNWAY ELEVATION

1. 1600 RPM or 16 inches MP.
2. Gear down and flaps 1/2 - approx.

<u>STEP 2</u> - AT 500 FT. POINT (500 FT. ABOVE THE RUNWAY ELEVATION)

1. Flaps set for approach and landing.
2. Airspeed stabilized at 1.3 x stall speed for aircraft wt. and flap setting.
3. Plan touchdown point as appropriate.
4. Landing check list complete.
5. Monitor rate of descent.

<u>STEP 3</u> - AT THE KEY POINT (300-350 FT. ABOVE THE RUNWAY ELEVATION)

1. Distance from touchdown approximately 1 mile.
2. Plan touchdown point as appropriate.
3. Monitor rate of descent.
4. Adjust airspeed and descent rate - as needed.

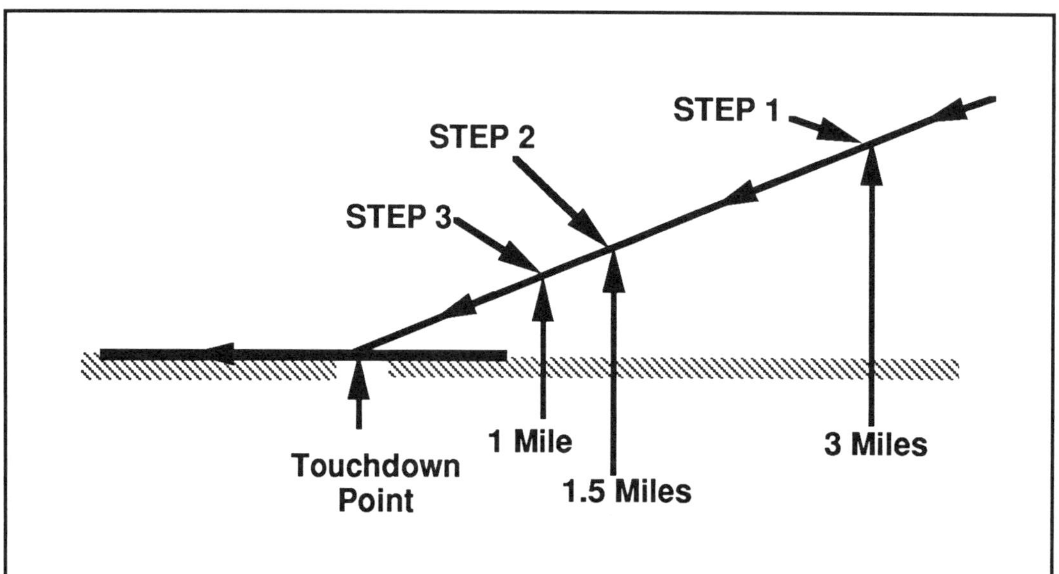

Figure 8-2

APPROACH PLANNING CHART
(Refer to Figures 8-1 and 8-2)

Pilot Name ____________________

Airplane Make and Model ____________________

N ____________________ Date ____________________

	STEP 1	STEP 2	STEP 3
POWER*	________	________	________
GEAR	DOWN	DOWN	DOWN
FLAPS	________	________	________
DESCENT*	______ FPM	______ FPM	______ FPM
IAS	______ KTS	______ KTS	______ KTS

* Make several approaches and fill in the blanks after you have refined your approach plan. You may want to make several copies of this chart.

Figure 8-3

APPROACH PLANNING CHART

For all approaches day or night from 1,000 ft. above runway elevation to touchdown.

CESSNA 172

	STEP 1	STEP 2	STEP 3
POWER	1500 RPM	1400-1600 RPM	1300 -1600 RPM
GEAR	DOWN	DOWN	DOWN
FLAPS	20 DEGREES	20-30 DEGREES	20-30 DEGREES
DESCENT	450-500 FPM	400-450 FPM	400 FPM (APPROX.)
IAS	SLOWING TO 65 KTS	65 KTS (APPROX.)	65 KTS (APPROX.)

BONANZAS - V-35, F-33A, A-36

	STEP 1	STEP 2	STEP 3
POWER	16 INCHES M.P.	16 INCHES M.P.	16 INCHES M.P.
GEAR	DOWN	DOWN	DOWN
FLAPS	1/2	FULL	FULL
DESCENT	600 FPM	500 FPM	450 FPM (APPROX.)
IAS	SLOWING TO 85 KTS	85 KTS (APPROX.)	85 KTS (APPROX.)

NOTE: These charts are for purposes of illustration only! In actual practice, the numbers can be expected to vary to meet the various operating conditions and aircraft weights.

Figure 8-4

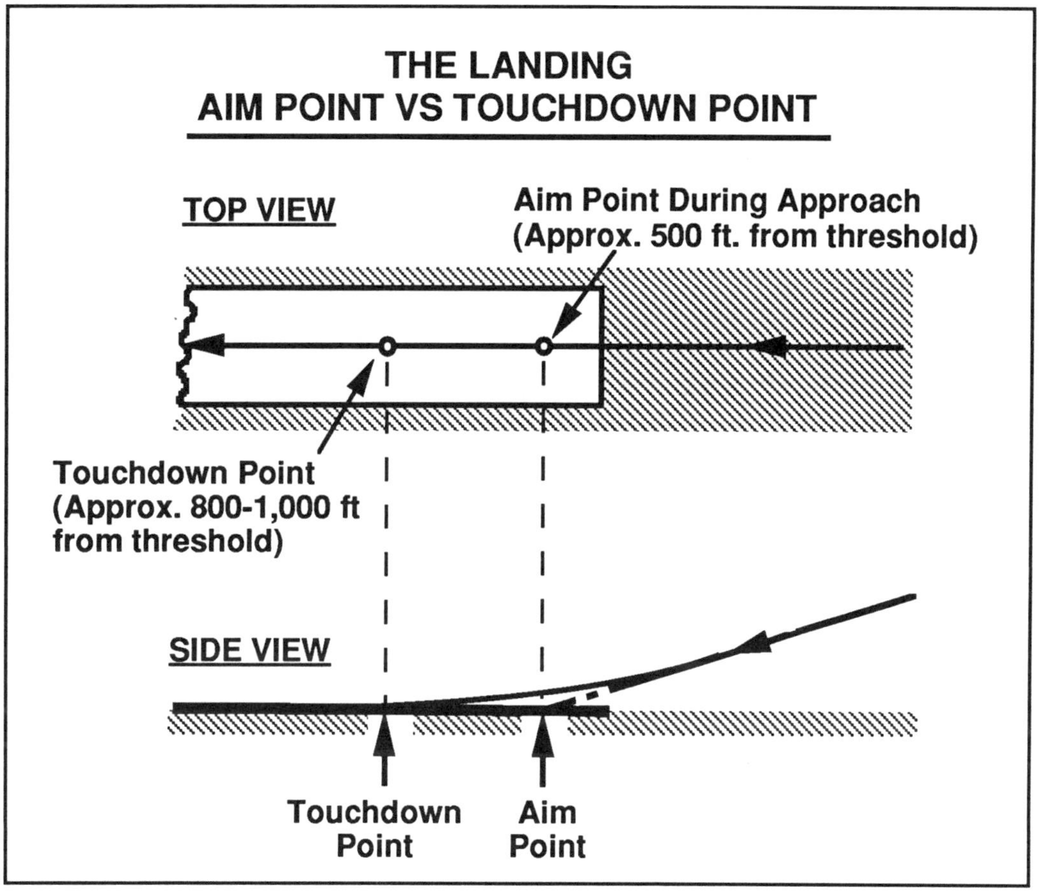
THE LANDING
AIM POINT VS TOUCHDOWN POINT
TOP VIEW
Aim Point During Approach
(Approx. 500 ft. from threshold)
Touchdown Point
(Approx. 800-1,000 ft
from threshold)
SIDE VIEW
Touchdown
Point
Aim
Point

Figure 8-5

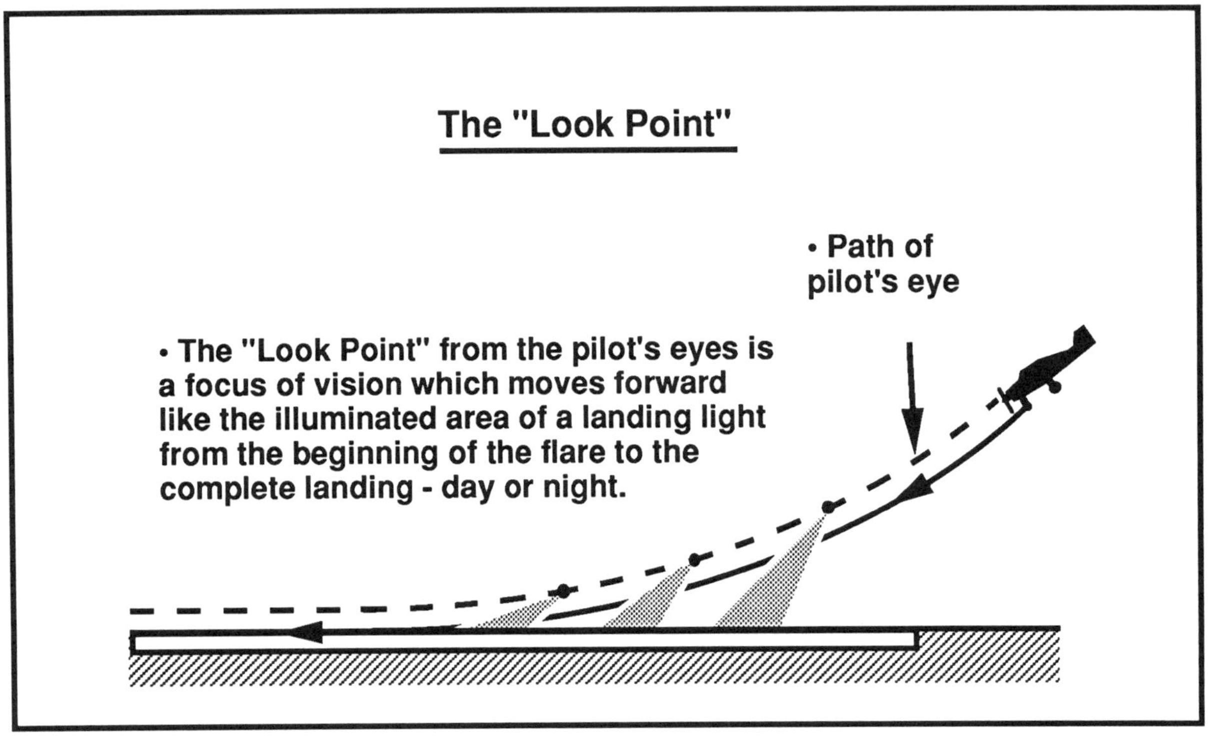
The "Look Point"
• Path of pilot's eye
• The "Look Point" from the pilot's eyes is a focus of vision which moves forward like the illuminated area of a landing light from the beginning of the flare to the complete landing - day or night.

Figure 8-6

LET THE MYSTERY BE EXPOSED

There appears to be a perception of mystery accompanied by frustration and great anxiety when a go-around/missed approach is either discussed or executed. This unease and the hazards associated with these operations can be disposed of very easily. We simply need to understand what we are asking the aircraft to accomplish and have a step-by-step plan for the overall operation.

We need to differentiate between go-around and missed approach. A *"go-around"* will refer to visual operations. A *"missed approach"* will refer to a situation during IFR operations wherein the runway environment is not visible to the pilot at the missed approach point specified in the instrument approach procedure. In this situation a missed approach is mandatory. Additionally, during IFR operations, a missed approach may be initiated at any time during an IFR approach for a variety of reasons.

> REGARDLESS OF THE SITUATION DURING APPROACH OR LANDING WHICH PROMPTS A PERSON TO **WANT OUT OF THERE-NOW,** THE PLAN AND PROCEDURE IS THE SAME WHETHER DEFINED AS A GO-AROUND OR A MISSED APPROACH.

THE PROCEDURE IS AS FOLLOWS:

POWER AND SLIGHT NOSE UP PITCH - applied simultaneously. I refer to this step as the *"TWO P's"* -- POWER AND PITCH. Apply takeoff power while raising the nose to 5 degrees to 8 degrees nose up pitch attitude. This action will set up a shallow climb and prevent any further descent onto the runway or into terrain or obstructions.

VERIFY THAT YOU HAVE IAS TO BEGIN FLAP RETRACTION

FLAPS - RETRACT TO APPROXIMATELY THE ONE HALF POSITION. This flap setting will vary amongst the many makes and models of aircraft. Consult the POH. The new flap setting will immediately dispose of much of the drag thus some acceleration will usually be noted at this point.

VERIFY THAT YOU HAVE THE IAS TO CONTINUE FLAP RETRACTION.

FLAPS - RETRACT - TO ZERO. Check the rate of climb instrument and assure that it indicates a POSITIVE RATE OF CLIMB.

VERIFY A POSITIVE RATE OF CLIMB

GEAR - RETRACT

RESUME NORMAL CLIMB - Adjust pitch, airspeed and elevator trim for V_y climb speed. See Figure 9-1. NOTE: This procedure parallels airline operating practices!

DISCUSSION

In my opinion it is incorrect to begin the go-around or missed approach by applying takeoff power and immediately retracting the gear. There is no need to be in a big hurry here. In fact, trying to hurry too much is a major cause of crashes during go-around attempts. No matter how many controls and switches need to be activated -- the first priority **ALWAYS** is to fly the airplane safely. Secondly, there is no wisdom in ever operating controls or switches faster than the brain can easily and accurately comprehend what is happening.

LET ME EXPLAIN WHY THERE IS GOOD REASON **NOT** TO RETRACT THE GEAR AS A GO-AROUND/MISSED APPROACH IS BEGUN. ACCIDENT REPORTS, PERTAINING TO GENERAL AVIATION AIRCRAFT, REVEAL THAT MANY AIRCRAFT ARE ALLOWED TO SETTLE ONTO THE RUNWAY AS THE GO-AROUND/MISSED APPROACH IS BEGUN. THE PRACTICE OF **POWER AND NOSE-UP PITCH** WILL PREVENT THIS PROBLEM.

When an unexpected touchdown involves a fixed gear aircraft or a retractable gear aircraft, with the gear still extended, the result will usually be no more than a rough and embarrassing bounce! However, when a retractable gear aircraft settles to the runway with the gear in transit toward the up position, the scene turns nasty. What is the time required to retract or extend the gear on a general aviation aircraft? The amount of time varies widely depending on the make and model of aircraft. Most gear retract/extend times fall in the 6 second to 10 second time span. Check this time on any retractable gear aircraft you fly.

Is there a pressing need to retract the gear in order for the aircraft to perform a go-around? If the aircraft is a fixed gear airplane the go-around will be accomplished easily if the flaps are retracted in a timely manner. If a retractable gear airplane can takeoff with the gear down it can go around with the gear down provided flap drag is reduced by a sufficient amount. So it follows that with any retractable gear aircraft the flaps should be retracted first as in the case of a fixed gear aircraft. A further reason to retract the flaps first is revealed when we reflect on the fact that **flap retraction is almost instantaneous** whereas gear retraction requires several seconds of time. An additional fact is that the amount of drag produced by full flaps extended will equal, or in some aircraft, exceed the drag of the extended landing gear.

Logic thus dictates that for **quick reduction of drag** the best action is to first retract flaps. Check the POH for your aircraft.

Aircraft handbooks for the Beech Duchess, Cessna Centurion, Mooney 201 and Cessna 414 are typical in that flap retraction is specified as being accomplished **before** gear retraction. In a few handbooks, there is no procedure for a go-around. In rare instances, mostly older light twins, gear up is specified to be accomplished before flaps are retracted. I like a note in the Cessna 414 handbook which states -- *"Do not retract the landing gear if another approach is to be conducted"* -- this means following a go-around. The reason is that many gear up landings occur following a go-around. The attendant excitement of a go-around so distracts the aviator that the gear down and landing check list is overlooked during the next landing approach. The result is then likely to be a gear up landing.

So my concluding words of wisdom are to be deliberate and take it easy while conducting a go-around or missed approach. The basic framework of the procedure is simple and as follows: Power and nose up pitch, flaps retracted to about one half, then flaps up to zero. If operating a retract gear machine, gear up and you are on your way.

When a go-around is initiated because of a traffic conflict with an aircraft ahead of you either on the runway or airborne, be sure to maneuver your aircraft so that the other aircraft is kept in sight. Communicate your go-around and further intentions so that the other pilots, and where applicable the controllers, will know your plans.

Do not be a loose missile in the traffic pattern -- promptly announce your plans for getting back on the runway or for diverting to another airport.

A suggested practice routine is to get in a roomy practice area at 1500 feet AGL or more and establish a descent of 500 FPM, in landing configuration -- at the proper final approach speed. Decide on an altitude you have selected to represent a runway elevation. As the altimeter gets to the *"100 feet above"* reading initiate a practice go-around. Repeat this exercise until it can be accomplished with no loss of altitude. When you feel comfortable, begin practice at the airport. Initiate a go-around at 100 feet above the runway then 50 feet above, 20 feet above, etc. until you can do a smooth and orderly go-around from just above the runway. If you feel any insecurity practicing go-arounds on your own, then by all means get a good CFI to go with you and assist you as is appropriate. Good Luck.

COULD THESE ACCIDENTS HAVE BEEN PREVENTED?

A light twin was too high on the approach for a night landing. Finding himself still airborne too far down the runway for a safe landing, the pilot initiated a go-around by adding power and immediately moving the landing gear control to the up position. The airplane continued descent until both props contacted the runway. The pilot then raised the nose and proceeded to climb out and fly the pattern with the props badly bent. On the next approach, the pilot forgot to extend the gear due to stress and excitement of the moment. The aircraft landed on the belly and off the end of the runway. The plane was destroyed and the pilot injured. This accident was easily preventable.

A student pilot was practicing full flap landings in a Cessna 150. He attempted a go-around but the flaps were never retracted from the full flap position. The drag with the full flaps was enormous. The airplane stayed airborne only briefly even with full power applied. It then stalled when about 30 feet above the runway and *"fell like a rock"*. The student pilot was slightly injured. The airplane was damaged beyond repair. This accident could have been prevented. In fact, the CFI had specifically told the student to make all solo landings with either zero flaps or 20 degrees of flaps until more approach and landing proficiency had been attained. In addition, the student was instructed **not** to use full flaps on solo landings.

A Typical Procedure for Go-around/Missed Approach

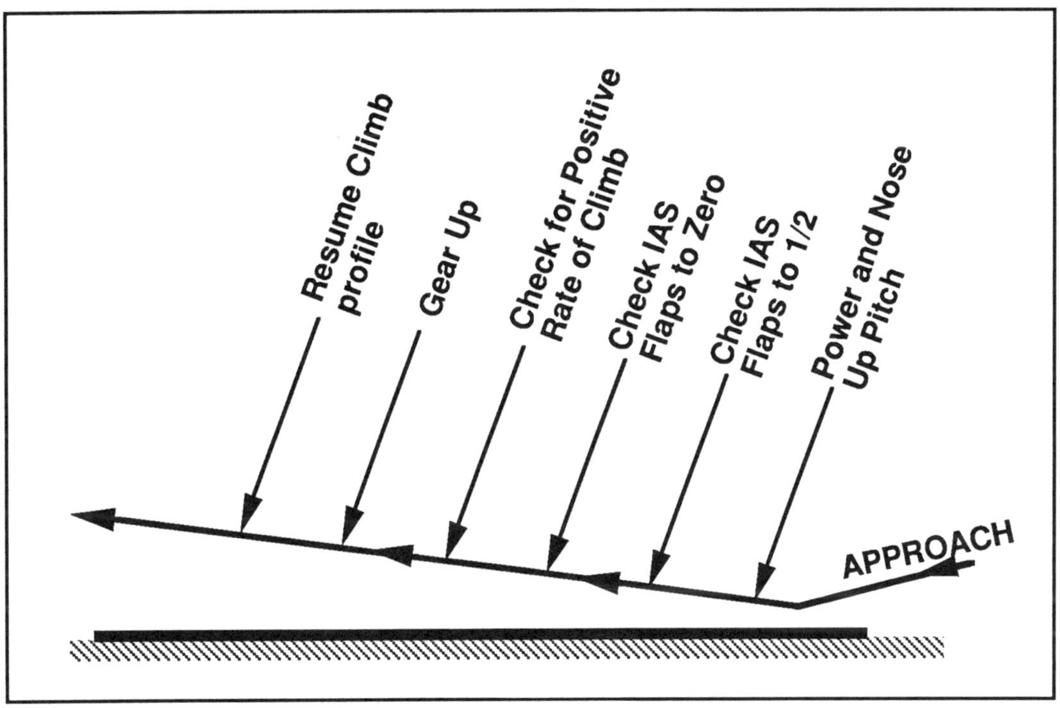

Figure 9-1

CAUSES OF BAD LANDINGS

Bad landing recovery techniques are a necessary and important part of each pilots bag of tricks.

Bad landings generally result from one of the following situations addressed in this chapter:

FIRST- Leveling off too high above the runway with a resultant abrupt drop onto the runway.

SECOND- Inadvertently flying onto the runway before the landing flare is completed -- with a resulting bounce into the air.

THIRD- Loss of directional control during any part of the landing operation with principal emphasis on the roll out following touchdown.

BAD LANDING RECOVERY ACTIONS ARE IN ORDER ANY TIME THE OPERATING SITUATION IS OTHER THAN NORMAL FROM THE BEGINNING OF THE LANDING FLARE UNTIL THE AIRCRAFT IS SAFELY CLEAR OF THE LANDING RUNWAY.

DISCUSSION AND RECOVERY PROCEDURES

FIRST- Let us consider the situation wherein the aircraft is **leveled off too high** above the runway at the end of the landing flare.

"Too high" means that the main gear tires are approximately 12 inches **or more** above the runway. If this excessive height above the runway is inadvertently maintained, as airspeed decreases and lift is lost, the result will be a bone jarring drop onto the runway. When the airplane drops abruptly onto the runway from a height of more than several inches the result may be damage to the landing gear and support structure.

This situation is usually caused by looking **too far ahead of the aircraft** or momentary inattentiveness during the landing operation. Review again the discussion entitled HOW TO LAND USING THE **LOOK POINT** in Chapter 8. Here is the procedure for recovery from the *"level off too high"* situation.

Apply power immediately. Make the power application which has a sound similar to the sound of the engine during run up. In extreme situations full throttle may be the appropriate initial application. Maintain directional control and a slight nose up pitch altitude throughout this recovery sequence. The objective is to momentarily keep the aircraft airborne just above the runway until the situation is stabilized. When the situation is stabilized, usually in 1 or 2 seconds, begin a smooth and steady reduction of power while maintaining the nose up pitch attitude and directional control. This power reduction will cause the aircraft to settle onto the runway in a smooth touchdown. Follow on with good directional control and the application of brakes as is appropriate. See FIGURE 10-1.

SECOND- This situation involves the airplane being **inadvertently flown onto the runway** before the landing flare is completed with the resulting bounce back into the air. This undesirable event is usually caused by looking at the runway environment **too close ahead of the aircraft** or by momentary inattention to the landing operation. See FIGURE 10-1.

The recovery procedure, for this situation, is the same as for the level off high situation. Here the airplane will be bounced back into the air. At the beginning of each of these recoveries the aircraft is in about the same situation i.e. airborne just above the runway, with a nose up pitch attitude and only slightly above stall speed. See FIGURE 10-1.

The recovery technique which fits this situation is repeated: Apply power immediately, maintain slight nose up pitch attitude and directional control. When stabilized, reduce power smoothly. The aircraft will settle onto the runway in an orderly manner. If a touchdown involves only a light *"skip"* then no recovery action is needed. Merely maintain the nose up pitch until the *"skip"* landing is completed.

A variation of this bad landing is when the aircraft goes through a series of moderate to fairly high skips after touchdown. This is called a *"PORPOISE"* landing which is caused by excessive airspeed at touchdown and when the aircraft is not in the normal nose up landing attitude.

The word porpoise is truly descriptive! Recovery action is to adjust the nose up to the landing pitch attitude and proceed with the bad landing recovery. In severe situations the best recovery may be to immediately add full power, lift the airplane off the runway and proceed with a go-around. Maintain directional control by aggressive application of rudder.

Keep in mind that these bad landing recovery actions will be moving the aircraft down the runway at a rapid pace. Thus a recovery action conducted on a short runway, or following a landing which is too far down the runway, may result in running off the end of the runway before the aircraft can be stopped. Maintain an awareness of the amount of runway remaining. To safely accomplish a bad landing recovery the remaining runway length must be sufficient for the roll out after touchdown.

THIRD- A most undesirable situation is the **loss of directional control** following any touchdown or bad landing recovery.

Most runway excursions involve inattention to the task of stabilizing and maintaining directional control throughout the landing roll up to and including the turn to exit the landing runway. Timely application of rudder control is necessary throughout the landing roll out. Be aggressive with rudder inputs when necessary.

The pilot should keep a hand on the throttle(s) throughout the landing and roll out. **Do not** retract the wing flaps, open cowl flaps, or turn off the transponder during the landing roll out. The momentary distraction of these activities have caused many pilots to lose directional control at a critical point. Additionally, these actions do not need to be accomplished during the roll out. Many instructors advocate retracting the flaps immediately after touchdown to transfer the weight to the wheels so that heavier braking can be applied. I consider this line of thinking to be more of an opinion passed from one aviator to another rather than a proven fact. I know of no proof that the landing roll is shortened by quickly retracting the flaps after touchdown followed by braking versus leaving the flaps extended followed by braking. The energy that is dissipated at touchdown combined with the friction of the tires on the runway will slow the aircraft to well below stall speed very quickly.

A tricycle gear airplane has a tendency to **WHEELBARROW** after touchdown if the touchdown is at excessive speed. In this situation the nose wheel is on the runway while the main wheels are still airborne. This can result in the loss of directional control.

When operating on slippery runways or in any crosswind situation the pilot should use aileron and rudder controls as appropriate to virtually *"fly"* the aircraft to a stop after any landing.

Keep in mind that a small burst of power during the landing roll out will help to regain and/or maintain directional control in any single engine aircraft. In a twin engine aircraft the use of differential thrust, properly applied, will aid in regaining or maintaining directional control during a landing roll out in very challenging crosswind situations.

CAUTION: Bad landing recoveries should be reviewed and practiced periodically under the supervision of a highly experienced CFI.

COULD THESE ACCIDENTS HAVE BEEN PREVENTED?
A Cessna 182 bounced during landing and swerved off the runway striking a concrete culvert and a drainage ditch. Witnesses said they heard no application of power for recovery from the bounce and noticed no attempt to accomplish a go-around. The pilot was killed and aircraft was destroyed.

A Piper Warrior flown by a student pilot collided with a fence while landing. During the landing the airplane porpoised several times then ran off the right side of the runway where it hit the fence. The airplane was substantially damaged but the student pilot was uninjured. The loss of directional control was a prominent factor in this accident. There was no reported attempt at correcting the porpoising.

FINAL THOUGHTS
When a landing situation gets to be really nasty, the best way out is full power and aggressive use of rudder to maintain and/or regain directional control while getting the aircraft airborne again. Take a deep breath and come around for another landing.

If you are not confident of your current capability at doing bad landing recoveries -- **you need help now!** Contact a good CFI and get some dual practice. Do not allow the airplane to self destruct because of your lack of bad landing recovery skills.

MOST LANDING ACCIDENTS OCCUR IN SITUATIONS WHERE NO RECOVERY ACTION IS TAKEN. LANDING SAFETY DEMANDS THAT EVERY PILOT BE PREPARED TO EXECUTE AN APPROPRIATE BAD LANDING RECOVERY WHENEVER THE NEED ARISES.

BAD LANDING RECOVERY

A single bad landing recovery technique is applicable whether the bad landing results from a rough and unexpected bounce - or - when the airplane has been inadvertently leveled off more than several inches above the runway.

At the instant that recovery is initiated, the two situations are similar in that the airspeed is slightly above stall, height above the runway is excessive and the pitch attitude is slightly above level flight.

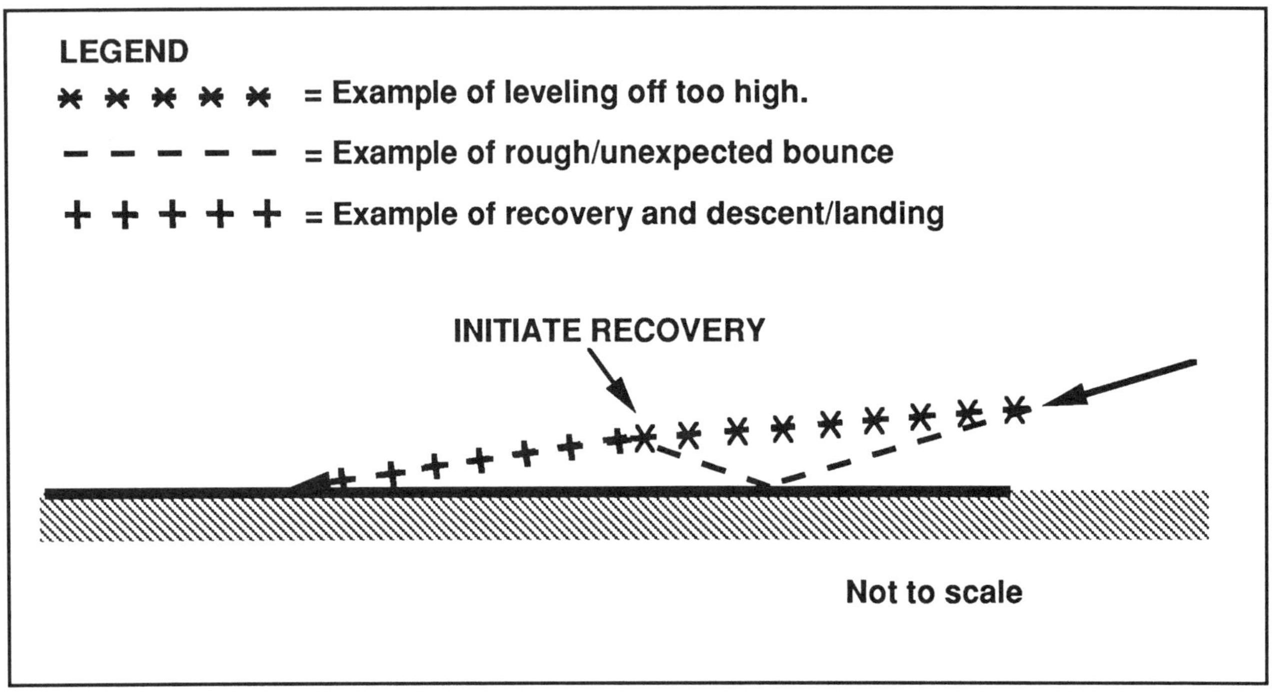

The sequence for a bad landing recovery is:

- Promptly add power (The sound level similar to a mag. check)
- Maintain directional control and pitch attitude
- As soon as stabilized - reduce power smoothly
- The airplane will begin to settle toward the runway
- Complete a normal landing

Figure 10-1

SAFE NIGHT FLYING REQUIRES EXTRA TRAINING

NIGHT FLYING IS MORE THAN DAYTIME OPERATIONS CARRIED OUT DURING THE HOURS OF DARKNESS. SPECIAL TRAINING IS REQUIRED IF SAFETY IS TO BE ASSURED.

We will now focus on each segment of night flying operations to highlight areas needing special emphasis.

NIGHT PRE-FLIGHT INSPECTION

This segment must be accomplished in a thorough and unhurried manner. Extra time and attention must be directed to each item on the check list. It may be necessary to repeat the check of oil quantity and fuel drained from the sumps. Personal injury must be guarded against throughout this activity. Do not be rushed. Keep every check on a definite basis. Do not take any short cuts. Follow the checklist. Use a good flashlight.

NIGHT ENGINE STARTING AND TAXIING

Extra vigilance is required to avoid hitting personnel, other aircraft, ground equipment and obstructions during this phase of operation. If the night airport scene appears confusing ask the ground controller for directions while you taxi outbound or inbound. These folks are always helpful but you've got to ask. If you have the IFR approach chart at hand study the taxi strip and runway layout prior to starting the engine(s). While taxiing take note of the headings. For example, if the taxi clearance is to runway 27 for takeoff then the parallel taxi strip heading should be approximately 090 degrees. If you are on a different heading you are not where you should be. Ask for assistance. Be extremely careful to comply with the taxi clearance when crossing active runways.

White lights indicate runway edges while blue lights indicate taxi strips -- however on a strange airport the sea of lights can still be confusing. **The great danger in night taxiing is to inadvertently wander onto an active runway!** This is extremely dangerous. Know your location at all times. Do not take chances regarding your location on the field.

NIGHT RUN UP

During the runup it is helpful to have a landing light turned on so that any inadvertent movement of the aircraft will be detected and stopped immediately. Use the flashlight as necessary to get a good look at every instrument, switch and control. I know of no general aviation single engine or light twin wherein the aircraft cockpit lighting is fully adequate for all night operations.

NIGHT TAKEOFF AND INITIAL CLIMB

Check the aircraft heading as the aircraft becomes aligned with the runway centerline. If there is a discrepancy of more than a few degrees -- check it out before you begin the takeoff roll. You could be on the wrong runway. When ready to roll, proceed as with a day takeoff thru lift off, at which point transition should be mode to the attitude indicator. Maintain wings level, runway heading and best rate of climb speed to 1,000 feet above airport elevation or higher if advisable. Pitch attitudes will vary from about 8 degrees nose up to 10 degrees nose up for most planes powered by reciprocating engines. Remember the pitch attitude which is appropriate to your plane and use it every time in takeoff and climb operations. The climb out should be by reference to the flight instruments along with all available visual references.

Major hazards are departing a lighted airport over a **black hole** or unlighted area. Here the flight instrument reference may be a life saver. NOTE: Use the same takeoff profile as for the day takeoff.

You do not need to have an instrument rating to utilize the flight instruments during VFR operations! Make use of every instrument which is available to you -- day or night! Flying into the ground after lift off is a definite possibility unless the appropriate nose up pitch attitude is maintained. Night departures, from airports surrounded by hills, tall buildings, bridges, and tall towers will require continuous best rate of climb performance in order to clear all obstacles by a safe margin. Cockpit lights should be re-adjusted if necessary and landing lights should be turned off as soon after lift off as is operationally convenient and safe.

NIGHT CRUISE CLIMB, CRUISE AND DESCENT

During this phase of flight extra attention is required throughout. Navigation may at first seem easier at night because the lights of towns and cities are bright. The downside is that towns and cities are more difficult to identify and distances from the aircraft to an upcoming town will appear nearer than you estimate. The town that appears to be 10 miles ahead may be 20, 30 or more miles away! So what can we do to navigate safely at night? Here again dead reckoning should be utilized -- namely, what is the heading and time between visual checkpoints and/or navigational fixes. The need for a current flight log is of special importance for night cross country operations. If the flight is proceeding via airways during VFR night conditions then the navigation is easier but the flight log is still necessary. When utilizing LORAN, RNAV, GPS etc. it is of the utmost importance to keep track of your position.

Otherwise, in case of an electrical failure or navigation equipment failure, you may not be able to proceed to a safe landing. Keep in mind that even a new aircraft and/or new navigational equipment can and does fail at very inconvenient times. These failures at night are scary and can end in disaster unless the pilot is prepared to cope with the failure. Night VFR cross-country will be much safer if flown on an IFR flight plan if the pilot is instrument rated.

NIGHT APPROACH, LANDING AND GO-AROUND

Special advanced planning of altitude and airspeed is required for a safe night VFR approach. Having identified the airport, the descent should continue to no lower than field elevation **plus 1,000 feet** until 3 miles from touch down. This may be a straight in approach or if in the traffic pattern this 3 mile point is approximately opposite the touch down spot while on the downwind leg. If the airport is in a valley, or surrounded by tall obstructions an appropriate **higher altitude** must be maintained until very near the airport. The pilot is responsible for knowing the minimum safe altitude along the ground track to the airport. IFR approach charts and enroute charts are a good source of minimum altitude information. Additionally, careful study of the sectional chart will provide the pilot with the elevations of obstructions and terrain in the area. Determine the safe operating altitudes in the vicinity of the destination airport prior to departure. Do not blindly fly toward the destination with no attention to obstruction clearance altitudes. Attempting to judge height above terrain or obstructions *"by eyeball"* is foolish and deadly -- **THE ALTIMETER MUST BE UTILIZED.**

NIGHT PRECISION VISUAL APPROACH -- (IDENTICAL TO A DAY APPROACH)

Keep in mind the importance of monitoring the altimeter readings and the rate of descent throughout the approach.

The descent rate will vary from about 350 to 500 FPM for most fixed gear singles and about 400 to 600 FPM for light twins and some high performance singles.

The rate of descent to maintain a 3 degree descent profile at a ground speed of 60 knots is 320 FPM, for 75 knots ground speed the descent rate is 395 FPM, for 90 knots ground speed the descent rate is 480 FPM, and for 105 knots ground speed the rate is 555 FPM. Pick out the rate of descent that is nearest the ground speed for most of your approaches and establish a ballpark descent range figure. For example an approach being flown using an indicated airspeed of 75 knots with a 10 knot headwind would need an average descent rate of about 350 FPM. Thus, during this approach the rate of descent range would logically be 300 to 400 FPM. If the descent rate is 200 FPM you should expect to be too high on the descent profile. If the rate is 500 to 600 FPM you will be dangerously low in a matter of seconds. Do not try to fly the approach by the rate of descent. Instead, use the instrument as a check or progress report. If an unwanted rate of descent does develop it can easily go undetected by the seat of the pants, whereas the rate of climb instrument would have provided the pilot with an early warning.

For many years I have used a final altitude safety check to avoid flying into the ground short of the runway which is a major cause of fatal accidents during night approaches. The plan is a follows: add 100 feet to the field elevation figure and keep that in mind as the altimeter reading which you will use as the visual minimum descent altitude unless the aircraft is positioned to cross the runway threshold at the normal height. Additionally the landing lights will be illuminating the pre-threshold and threshold area beginning at around 150 feet AGL.

This scheme has served me well especially during operations into unfamiliar airports at night. I encourage you to use this as a final check during all night landings. In all approaches, day or night, utilize the old reliable scheme of watching the **aim point**. If it moves down in the windshield you are getting too high on the approach. If the aim point moves up you are getting low on the approach profile and immediate corrective action is required. There are various proposals about watching the angle of the runway lights, the runway slope, etc. All of these, are in my opinion totally worthless. The PRECISION VISUAL APPROACH will get you to over the runway threshold in position to land on or near the desired touchdown point. Keep in mind that we are referring to operations involving regularly used airports where runway slopes are minimal. Most runways do have some slope gradient but that will not impact on normal landing operations.

FOR NIGHT LANDINGS THE PILOT SHOULD UTILIZE THE SAME LOOK POINT AS PREVIOUSLY DISCUSSED FOR DAY LANDINGS.

It is a convenient coincidence that the landing light or lights will illuminate the prescribed look point area if a normal tail low altitude is attained during the night landing.

LANDING WITHOUT RUNWAY EDGE LIGHTS

For operations into airports, where there are no runway edge lights, extra caution and attention are required. Fly the approach as previously prescribed. There are differences to be taken into account when there are no runway edge lights. First - approach alignment is accomplished by keeping the far end lights aligned directly above the threshold lights.

At approximately 150 feet AGL the landing light or lights will illuminate the pre-threshold area the same as if runway edge lights were illuminated. Second - the landing flare and the touchdown must be accomplished without the benefit of all the peripheral visual cues associated with edge lights.

The **LOOK POINT** continues to be the main reference for the landing. The roll out will be normal if **runway edge lines** and or centerline markings are plainly visible. If there are no visible runway or taxi strip guidelines, the landing, the roll out and all taxiing operations will require extra caution. In most situations unlighted runway edges can be recognized by contrasts of textures such as grass or weeds against the runway edge. Contrasts in color may be helpful such as the runway color against snow, sand or vegetation which may be brown or green. Snow which has been plowed to the edge of the runway presents a definite view of the useable runway boundary.

HOW ABOUT LANDING ON AN UNLIGHTED RUNWAY? DO NOT EVEN CONSIDER THIS EXCEPT IN A DESPERATE EMERGENCY.

NIGHT LANDINGS WITH INOPERATIVE LANDING LIGHT(S)

On occasion it may be necessary to land without the benefit of landing lights. These situations result from an enroute electrical system failure or a burned out landing light unit. If you encounter this situation avoid attempting a landing on a runway which does not have operable runway edge lights. A preferred course of action would be to divert to an airport which has operable runway edge lights.

BAD LANDINGS RECOVERIES - DURING NIGHT OPERATIONS

For night bad landing recoveries utilize the same techniques as for day bad landing recoveries. It has been my experience that the most likely error in any night landing with or without landing lights is that the aircraft is leveled off too high. To guard against this happening -- begin the flare in a normal manner. If touchdown does not occur within about 4 to 6 seconds -- add a small amount of power (approximately 1,000 RPM) and allow the plane to gently settle to the runway. The added power will be just enough to slow the decrease in airspeed while you feel for the runway. If the touchdown is smooth you have lucked out. If there is a **slight bounce** no corrective action is necessary as the aircraft will immediately touchdown again after a short skip. Do not relax the elevator control in this situation. If the touchdown is rough enough that the airplane is bounced back into the air more than a few inches -- immediately add a small burst of power. This should sound similar to the power for a run up. This will instantly stabilize the aircraft above the runway at slightly above stall speed. Now while holding the pitch attitude at or near landing attitude -- reduce the RPM smoothly. The aircraft will then settle back to the runway with a reasonable touchdown. It is important to maintain directional control throughout this event. Close the throttle at touchdown and proceed with a normal landing roll out. Consider an immediate go-around if the outcome of recovery techniques appear in doubt. Note that bad landing recoveries at night are identical to daytime recoveries!

CAUTION ON THE USE OF LANDING LIGHTS

The use of landing lights may in some cases set up a hazardous situation. Let's explore examples applicable to both takeoffs and landings.

First I will relate a personal experience during a night takeoff from the Manhattan, KS. Municipal Airport in the early 1950's. The time was shortly after dark on a clear cool evening. I was flying a Lockheed 10-A (SHADES OF AMELIA EARHART) with a load of soldiers bound for Detroit City Airport. The visibility in the area was about 5 miles in light haze. The sky was clear and the wind was calm. I departed on runway 13 which took me over the river bottoms. At 300 feet AGL I began a turn to a heading of 075 degrees while struggling up to 500 feet AGL and accelerating to 120 MPH. This altitude and airspeed was a target point wherein it was rumored flight could be continued if one engine failed! Fortunately it was never my lot to test the validity of this rumor! I had grave doubts. Throughout this operation I had the landing lights on and I was very intent on watching the engine instruments, the altimeter and the airspeed indicator as I climbed through 400 feet AGL. The cockpit ceiling above my head contained a small eyebrow window some 12 inches square. At this point I became aware of a red light visible in my peripheral vision and coming through the eyebrow window above and left of my head. In a heartbeat my mind went into the *"WHAT THE HELL IS GOING ON"* mode! I flashed a quick look out the windshield and what I saw looked like a giant grey sheet which was caused by the landing lights shining against the haze! At this moment things were not going well. My location was now about 3 miles south of the Kansas State University complex. The two Pratt and Whitney 450's were still screaming at full bore. My survival instincts snapped my vision to the attitude indicator and what I saw was very disturbing! Nose up pitch was near 15 degrees. Normal nose up pitch is 7 degrees. The left wing was down in a bank of nearly 60 degrees and the airspeed was dropping thru 85 MPH! Faster than the speed of light I simultaneously rolled the wings level and lowered the nose to about 3 degrees nose up pitch in order to conserve altitude.

I turned off the landing lights and stared, statue like, at the instruments while I struggled the remainder of the way up to 500 feet AGL. Out of my left cockpit window I glanced at downtown Manhattan as we roared into the night. This whole episode occurred in approximately **five seconds!**

I was not a novice at night flying -- I had nearly 2,000 hours of night experience which included airline DC-3, Trans-Pacific ATC in C-54's and a variety of general aviation aircraft. Why did the episode happen? What went wrong? The answer is very simple. The landing lights were "blinding" me as they shown against the haze. Secondly, I was relying on feel of the aircraft while concentrating on the engine instruments, airspeed and altimeter. I had ignored the artificial horizon for too long -- maybe 5 to 8 seconds. What was that red light which I saw through my "eyebrow" window? It was the red obstruction light on top of a radio transmitter tower which was located on the KSU campus!

The safety message then is as follows: First, during takeoff turn off the landing light or lights soon after you cross the departure end of the runway. Secondly go immediately to scanning the flight instruments as during IFR climbouts. If there are lights on the ground in your departure area -- you have a lucky bonus. If you are climbing out in a **black hole** you will be proceeding safely. All turns should be made utilizing a 20 degree bank.

MORE ON LANDING LIGHT USEAGE

If any visibility restriction, such as haze, blowing dust, smoke, rain, snow or fog, cause the forward visibility to be reduced -- then consider leaving the landing lights off until on short final approach. These are **landing lights** and so called for good reason.

Do not confuse airline landing lights with general aviation landing lights. The typical airline aircraft electrical system generates enough electrical power to accommodate an apartment house. Thus the transport has very powerful landing lights. The typical general aviation aircraft electrical system is an approximate copy of an automobile system.

Avoid having the landing lights on during initial climb or landing approach if these lights will cause the pilots' forward visibility to be reduced significantly.

The use of landing lights for recognition and collision avoidance is good practice of course. It is the use of landing lights below 1,000 feet AGL which may present a hazard of pilot *"blinding"*. The use of landing lights is a judgement call for the P.I.C.

Landing lights should normally be on during the takeoff roll and through the start of initial climb. Additionally these lights should be on beginning at about 200 feet AGL on the approach -- unless the atmosphere is clear and visibility is good.

COULD THESE ACCIDENTS HAVE BEEN PREVENTED?

A Mooney M-20C was doing night touch and go landings at a Minnesota airport. While rolling along the runway, after touchdown, the pilot turned up the interior lighting so he could determine the position of the wing flaps. The pilot lost directional control. The aircraft ran off the runway and struck a snow bank. The aircraft was damaged substantially but the pilot, who was the only occupant, was uninjured. Night flying requires extra precautions because of the reduced visibility.

Solo night landings should be of the full stop variety. Solo night touch-and-go landings present many opportunities for distractions and mistakes due to the limitations of cockpit lighting.

A Piper PA-28 was on a night VFR flight. The pilot overflew the airport while attempting to activate the runway lights by keying the aircraft transmitter. The pilot stated that he wanted to observe the wind sock. He was unable to activate the runway lights. The pilot elected to land without the runway lights being illuminated. The aircraft touched down near the departure end of the runway. The aircraft bounced, then continued past the end of the runway. After the accident the pilot observed a tailwind component. The runway lights operated normally when keyed on the correct UNICOM frequency. The aircraft was substantially damaged. The passenger escaped injury while the pilot received minor injuries.

Several pilot actions should be discussed here. First, diverting to a nearby lighted airport would have been a good option. Secondly, when things don't work out with cockpit switches, especially avionics, the most likely cause is the **old demon -- COCKPIT TROUBLE.** In this instance a mistake may have been made in copying the UNICOM frequency. When the runway lights did not come on why wasn't the frequency questioned and rechecked? Thirdly, had a PRECISION VISUAL APPROACH been flown the chances for a touchdown near the approach end of the runway at the proper airspeed, would have been greatly enhanced. Attempting to land on an unlighted runway is unwise and extremely dangerous under any circumstances!

A good flashlight and extra attention to details are invaluable in night flying.

A Cessna 402 was on a VFR approach into an airport in Maine. The approach was being made at dusk. When near the runway a sudden sink rate developed. The pilot applied power and assumed climb pitch attitude, then felt a thud beneath the aircraft. Suspecting that the right main gear had been separated the pilot executed a go-around. On the subsequent landing the pilot held the right wing up as long as possible. As the aircraft slowed the wing dropped and the aircraft went off the runway. Investigation revealed gouges in the terrain, short of the runway, where the right main gear hit the ground and was torn from the aircraft. The pilot and passengers escaped injury.

THERE IS NO SUBSTITUTE FOR BEING ON SPEED AND ON DESCENT PROFILE UNTIL OVER THE RUNWAY THRESHOLD.

NOTES FOR SAFE NIGHT FLYING

Safe and reliable night flying requires extra discipline and preparation. The adequacy of cockpit lighting in most general aviation aircraft, is marginal. The pilot needs one and preferably two reliable flashlights readily available at all times. Special care and extra checking are needed when tuning avionics. All navigation and approach charts are more difficult to read at night. Maintaining the flight and fuel burn out log becomes a slow and demanding chore. Keep in mind that night flying is not just day operation with the lights turned on! Night flying is very serious business. Pilots with sloppy planning, poor navigating skills, poor decision making and lacking a good **PRECISION VISUAL APPROACH** plan, are ticking time bombs!

RECOMMENDED ADDITIONAL NIGHT TRAINING

The FAA minimum night time requirements are inadequate in my opinion. The following are my recommendations before you venture more than 50 miles from your home base at night.

NIGHT AIR WORK - DUAL - 1.5 HOURS

Climbs, descents, medium turns, approaches to stalls and recovery from unusual altitudes by use of instruments. If you do not have an instrument rating - so what! The procedure is very simple i.e. 1. Level the wings, 2. If the airspeed is higher than cruise reduce power -- if the speed is lower than cruise add maximum power, 3. Simultaneously bring the pitch to slightly above level flight and hold this pitch attitude using whatever elevator force is required! Readjust power as appropriate. The objective is to recover with minimum loss of altitude.

NIGHT LANDINGS - DUAL - 2.5 HOURS

4 full stop landings with 2 of these requiring bad landing recoveries. The CFI will know how to set up for a simulated bad landing.

2 full stop landings - without the use of landing lights.

2 cross wind landings - full stop.

1 go-around begun approximately 10 feet above runway.

NIGHT LANDINGS - SOLO - 1.0 HOUR (Under CFI supervision)

4 Full stop landings

NIGHT XC - DUAL - 5.0 HOURS

3 legs each to be a minimum of 150 NM in length. Included in this training flight should be approaches and landings at 4 airports along the route. Each of these airports should be ones at which the trainee has not previously made a night approach and landing.

The above course of additional night flight training and practice will enhance the safety and accuracy of future night flying operations.

Maintain night currency as required and/or desired. Night flying is very serious business. Be fully trained and competent before launching off into the dark of night!

THERE ARE NEEDLESS ACCIDENTS DURING IFR OPERATIONS

It is my opinion that a disproportionate number of incidents and accidents are associated with IFR operations in general aviation. As NTSB reports are studied, I am reminded that the same accident scenarios are repeated over and over -- year after year. What is wrong with the way we conduct our IFR operations? Is our instrument proficiency really adequate for the planned flight? Is our aircraft capable of safely completing the planned flight? Are we as pilots mentally and physically up to the task at the appointed time? Are our management and decision-making skills fully developed and utilized to assure that our IFR operations are safe? Keep in mind that accidents during IFR operations are very likely to result in death for all persons on board.

THE PILOT IN THE ROLE OF MANAGER

IFR OPERATIONS ARE VERY SERIOUS BUSINESS. PEOPLE ARE BEING KILLED DURING IFR OPERATIONS FOR MANY REASONS--PROMINENT AMONG THESE IS THE PILOT'S POOR DECISION-MAKING AND ENROUTE FLIGHT MANAGEMENT.

At this point I want to draw special attention to the fact that **the pilot must assume the role of manager during all operations.** In no flight operation is the pilot's role as manager more important and critical than in IFR operations. According to Webster's dictionary the word **manage** means *"TO HANDLE OR DIRECT WITH A DEGREE OF SKILL"*. It is unfortunate that the FAA examination requirements and thus the associated training are heavily biased toward stick and rudder proficiency plus some judgement references. So if the pilot is to be a manager, what does this responsibility entail? Read on!

PRACTICAL APPLICATIONS OF MANAGEMENT

During a weather briefing remember that a weather forecast is just that -- *a forecast*. A prediction of the weather conditions that are **likely to occur** at specified locations and times. Always remember that a forecast is not a guarantee! When very complex or unusual weather systems prevail the forecast may be little more than a best efforts guess.

The good manager, when receiving a weather briefing will assure that all appropriate weather and wind information is **received and understood!** Whether the weather briefing is received by means of high-tech equipment, directly from a briefer by phone or in person, some points need special attention. The following are points to keep in mind.

The standard weather briefing is not always standard. Ask for any added information which you need. The **icing level** must be understood as applied to the proposed route of flight. Are the safe operating altitudes expected to be free of ice? If not are there alternate altitudes that could be safely utilized? Are there recent PIREPS for this route? If not, can the FSS get PIREPS for you? Keep in mind that enroute **icing is a MAJOR KILLER** in IFR operations.

Temperature and Dew Point

These two items are key to understanding present weather situations and predicting likely future weather conditions. The good pilot/manager will ask for and copy the sequence reports for all stations along the route and the planned alternate airport. Then, while enroute, this good manager will get each updated station report and make appropriate comparisons.

In this scheme, when the weather shows any tendency to go unsafe the manager/pilot will be the first to know! This early warning allows a safe and professional operating plan to be devised and followed. On the other hand, if weather improvement is in progress, the pilot/manager can fly along with that relaxed and happy feeling. Each pilot should be capable of detecting and predicting weather trends with a reasonable degree of accuracy.

RED FLAGS

During IFR operations the good manager will associate a **RED FLAG** with any of the following situations.

When the temperature and dew point spread is 3 degrees F or less.

When a front is to be crossed enroute or when one is expected to be within 50 NM of the course during the airplanes travel through the area.

When the destination is near a significant body of water.

When precipitation is expected enroute or during the approach and landing.

When the destination runway is expected to be slippery.

When there are forecasts of TSMS, hail, icing, moderate or severe turbulence along the route.

When the surface wind at the destination is expected to be more than 20 knots.

When the destination or alternate is a one runway airport.

Each pilot can think of other significant **red flag** items. It would be a good idea to make a check list of the red flag items considered important. These red flags are indicators that extra caution and careful consideration are in order. Flight may well be conducted in a safe and orderly manner with some warning flags present.

However, red flags may also lead to a decision to delay or cancel the trip. Deciding to delay or cancel a flight, when appropriate to do so, is the mark of a truly safe and professional pilot. A sound and non-emotional go-no go decision is absolutely mandatory! This is a judgement call which is of the utmost importance.

Thunderstorms are a KILLER.
To operate in and through areas of forecast or active TSMS requires that the manager/pilot be getting continuous updates, via radio and visually, so that movement and intensity of these rumblers can be seen and forecast. TSM avoidance by the use of radar is the best practice. Operation in the area of active TSMS should involve slowing to the appropriate rough air IAS. Set the power. turn off the auto-pilot altitude hold function and monitor the weather situation as the active cells are detoured. At this point suitable alternate airports should be in mind continually. Diverting to an alternate airport along the route or even returning to the departure airport are all reasonable and professional courses of action. Keep a healthy respect for TSMS. Never knowingly fly near or into one of these monsters while on the gauges!

Let's now consider another **killer in IFR operations-FUEL EXHAUSTION.**

RUNNING OUT OF FUEL AT ANY TIME, REGARDLESS OF THE CIRCUMSTANCES, IS INEXCUSABLE AND INDICATES THE PILOT'S FAILURE AS A MANAGER.

I do not know of any reciprocating powered light twin or single engine aircraft which has a continuously reliable fuel quantity indicating system.

Some are good sometimes -- but in general, safety is best served when the pilot/manager maintains a skeptical view of what the fuel gauges read. The only 100% safe fuel management plan is to make a visual check of the fuel before engine start. This should be followed by realistic fuel burn out rates versus time on the clock. This is the only sure way to know at any point in time how much fuel has been burned and how much useable fuel remains.

FUEL GAUGE INCIDENTS

Take note of the following zingers which I have encountered with fuel gauges on general aviation airplanes.

> Inflight electrical failure which rendered the electrical fuel gauges inop.
>
> Fuel gauges wherein the needles stuck in some position and did not move further even though fuel was being consumed from the subject tank.

In 1991 my wife and I were flying a Piper Lance from Denver to St. Augustine, FL when a not too unusual event occurred. It was night IFR at the time as we were being vectored around the Jacksonville, FL area. When about 40 NM southwest of Jacksonville my wife said - *"both fuel gauges just went to zero."* She still believes in fuel gauges -- foolish lady! My fuel management plan was that I would have 45 gallons of fuel on board at the time of beginning the approach into St. Augustine. I immediately turned to *"you gotta fly the airplane as first priority"*. I called the ATC controller and told him I had a discrepancy in my fuel quantity indicating system. While I was not yet declaring an emergency, I was nonetheless requesting to know of any airports along my track that might be candidates for a deadstick landing in case that big old Lycoming ran out of food.

I further requested direct from present position to St. Augustine and declined his clearance to *"descend to and maintain 3,000 feet."* I also reduced power toward a more fuel efficient power setting and continued at 7,000 feet toward St. Augustine. These controllers are fantastic, but you have got to let them know what your situation is, what your intentions are and what assistance you need. Anyway, this gentleman stayed with me right up to the St. Augustine airport. Fortunately, the ceiling was 3,000 feet scattered as I approached the airport. We landed uneventfully and when the aircraft was fueled the next morning I calculated that we had landed with about 45 gallons on board.

After the narrative on how I managed the flight let me elaborate on other thoughts and possibilities which were involved. Approximately 30 minutes prior to this irregularity we had concluded about 1:15 of IFR, in heavy rain and occasional moderate turbulence, while in the area of Montgomery, AL. My first thought was that the electrical wiring had somehow gotten wet during our operation through the heavy rain. Really disturbing, however, was my concern that the fuel sump drain handle had been bumped open by some of the luggage we had loaded in that area. I rechecked the fuel invoice from Meridian, MS to verify the amount of fuel purchased there. Needless to say, my ever present flashlight was very useful that evening. When we departed St. Augustine two days later the fuel gauges had returned to normal and remained so all the way back to Denver. I still don't trust the gauges and don't rely on them. The basis of my fuel management is still the good old fuel log which reflects time and rate of fuel burn.

In another situation I was responding to a friends plea to fly a quick cargo charter from Denver to Gillette, WY.

It was during the winter of 1975 beginning about half an hour after dark. The Denver weather was nearly clear with a front forecast to arrive at Gillette half an hour after I planned to be on the ground. When I arrived at the airplane I made a hurried preflight -- bad idea! This included an inadequate check of fuel quantity -- very, very bad idea! The friend stated that there was plenty of fuel and further that a complete new fuel quantity indicating system had been installed because the original **had never worked right!** The gauges indicated each tank was half full. This was a red flag which blew right past me! The next red flag was that this was a new Cessna Turbo 210 with only 90 hours since it rolled out of the factory. New airplanes are to be approached with skepticism in my opinion -- but in the heat of battle this red flag sailed right past me too! I had already worked all day. I was tired and was missing dinner. Let's hesitate a minute and take stock of what I am allowing to happen.

I AM IN THE DEVELOPMENT OF AN ADVANCED CASE OF **GET-THERE-ITIS**. PURELY BY LUCK THE AILMENT BARELY MISSED BEING FATAL LESS THAN 1:30 LATER! READ ON!

I made a hurried VFR departure northbound via Cheyenne. The ground speed was great and the smell of the new airplane was very pleasant. I set up my flight and fuel log and settled back at 10,500 feet to roar along into the night. But wait -- something was strange here -- it was the fuel gauge readings! Both gauges were showing an increase in fuel quantity as I flew along! The Cessna 210 series are among my favorite airplanes and here I have one **making fuel as it flies!** Wait until I tell my friend about this! I bore on and verified that these fuel gauges were indeed unreliable! I am now worrying about whether or not my flashlight check of fuel at Denver was really adequate. The tanks were observed to be about half full. Very bad fuel quantity check!

I called Casper FSS for the weather and received the news that Gillette was expected to go IFR within 30 minutes with a 500 foot ceiling and blowing snow. I thought, I am only 15 minutes out so I can beat the weather. Soon the lights of the town were in sight and I started downhill and at the same time looked at my JEPP chart for the airport data. At about 2,500 feet AGL I became aware that the lights of the airport, which is north of town, were no longer visible. I decided to go on down a bit because there would certainly be a hole in the clouds and I could go straight in. Wrong! I had to give up and do a VOR approach. The next 12 to 15 minutes were spent in agony wondering what my fuel quantity was. Would the engine drink the last drop before I got on the runway? What kind of terrain was I over during the approach? This was my first time ever to be anywhere near Gillette much less the airport. My bad flight management had caught up with me and I did not like the situation. The weather had quickly gone minimums.

I turned to the important first priority of **always fly the airplane.** I wanted desperately to find the runway on the first approach as I doubted that I had enough fuel for a second attempt. I flew one of the tightest approaches of my career and thanks to a lot of good airline training, experience, and luck I got to the runway on the first approach. After parking I used a clean piece of wood to check fuel quantity. Was I out of fuel? Almost! I had no more than 2 gallons total for two tanks. That equates to about 7 minutes of flying time. I will always remember how that accumulation of bad flight management decisions came within minutes of causing a disaster! Never again!

GENERAL IFR RECOMMENDATIONS

The plain everyday job of flying the airplane can be a **KILLER** too. The following are points which I consider to be worth incorporating into every IFR pilots bag of tricks.

Check the compass and directional gyro readings as you line up on the departure runway. If there is a discrepancy between runway heading and directional instruments -- do not take off.

Before beginning the takeoff be sure to have on the yoke the IFR chart for the primary IFR approach to the departure airport. In the unlikely event that a problem arises after you have departed into the murk, you have only to advise ATC and request the specific IFR approach which you desire. The mental confusion and frantic pawing through papers and charts is eliminated, time is saved and you will look like a pro.

On climb out monitor the rate of climb and pitch altitude. Each year airplanes fly into the ground within one to five minutes after lift off. Keep it climbing. Up, up and away!

Avoid departing VFR to pick up the clearance in the air. This is very dangerous. NTSB reports contain many catastrophes caused by the pilot trying to maintain VFR while waiting for the IFR clearance. Flying into the ground, hills and all sorts of obstructions are common conclusions to this kind of operation. If I file IFR I wait on the ground until I get my IFR clearance - no exceptions! If this turns out to be a phone clearance with a void time -- so be it. No clearance -- no go, period!

While enroute manage the flight and review the IFR approach chart for the destination airport.

When being vectored to the final approach course and when turning to the intercept heading -- begin to slow down toward the approach speed and budget your time between the clearances being received and flying the airplane. This includes setting up the configuration and running the landing check list. All these items cannot be accomplished simultaneously so have a workable sequence that fits your operation. The area near the FAF is where operating errors and omissions frequently occur.

When inbound from the FAF, on a non-precision approach, fly a pre-determined descent rate. For most general aviation aircraft I use 700 FPM. The major error here is that the descent is begun a bit late and the rate of descent is too shallow. This may result in a pilot induced missed approach if the weather is actually at minimums. For an ILS know the expected "ballpark" descent rate before intercepting the glide slope. Be positive in establishing this rate of descent initially. Then, if high on the glide slope, correct by increasing the descent rate by 200 FPM. If low decrease the descent rate by 200 FPM until back on G.S.. Thereafter make the descent corrections smaller and smaller. At 1,000 ft. above the runway you should be pretty well locked on. A conveniently located sticky note will help in monitoring altitude during IFR approaches. Write the altimeter reading 1,000 ft. and 500 ft. above the runway/airport, as appropriate, and the MDA or DH.

Lastly, I consider the circling approach to be the most hazardous of the IFR operations. The only advantage to the circling approach is that the crash scene will be near the airport! Review the PRECISION VISUAL APPROACH again. Never allow the altimeter to read less than 300 feet above the airport elevation until you are lined up on final approach and within one mile of the runway. This rule may save your life!

I recommend adding 300 feet and one mile to all circling minimums unless you have a competent co-pilot on board and you are a pilot who has lots of hard IFR experience.

Do not allow the altimeter to read less than airport elevation plus 100 feet until you are coming up on the runway threshold. The greatest risk factor in a circling approach is flying into the ground some place along the circuit or during the final approach.

THE MAIN IFR KILLERS ARE:
- CONTINUING VFR FLIGHT INTO IFR CONDITIONS
- THUNDERSTORMS AND ICE
- LACK OF OVERALL MANAGEMENT BY P.I.C.
- FUEL EXHAUSTION WHILE ENROUTE
- POOR PILOTING DURING DEPARTURES
- POOR PILOTING DURING APPROACHES

Help prevent the above very serious problems which cause a large percentage of day and night IFR accidents. If we can eliminate these problems, lives will be saved, airplanes will be saved and untold amounts of grief will be prevented.

COULD THESE ACCIDENTS HAVE BEEN PREVENTED?

The pilot had landed in Florida for a vacation at which time he ordered the tanks topped. They were reportedly filled. Eight days later the pilot filed an IFR flight plan to Nashville, TN. Near Columbus, GA he changed destination to Huntsville, AL for fuel. The aircraft crashed about one mile short of the runway to Huntsville. The pilot was reported to have not made a visual fuel quantity check prior to departure. Probable causes were listed as inadequate pre-flight inspection, inadequate fuel consumption calculations and poor inflight planning.

A suitable caption might be *"if you don't have it in the tanks you can't burn it."* Aviation gas can be mixed with auto fuel -- thus the theft of one or two cans of gas from a tied down airplane is enough to top off the old hot rod! The theft of aviation fuel is more common than you might think! A ramp may be pretty dark after midnight and fuel theft can be accomplished without much trouble via wing or sump drains. I parked a Bonanza S-35 on the outer ramp at one of the busiest general aviation airports in the country. I estimated that I had 35 gallons on board at shutdown. The next forenoon I stuck the tanks and they were so near empty that I decided not to attempt taxiing to the FBO for fuel but called for the fuel truck instead! This thief stayed for the last drop!

In 1980 a Bonanza on a night IFR cross country flight in Texas crashed. There were two fatalities. During the weather briefing the FSS specialist reportedly failed to pass to the pilot a forecast of embedded TSMS along the route. Also reportedly not passed to the pilot were the contents of an AIRMET which called for low ceilings and visibilities along the planned route of flight.

In the state of Washington the pilot of a single engine aircraft was advised that VFR operation over his proposed route was not recommended due to low clouds and marginal visibility. He, nonetheless, decided to fly the route at night and encountered low clouds and marginal visibility. The aircraft crashed when it hit trees enroute. The aircraft was destroyed and pilot sustained injuries.

A pilot with instrument rating was enroute from Utah to the Denver area at night in IFR conditions, when misfortune struck. While cruising at 17,000 feet MSL the pilot reported ice build-up and requested a clearance to climb higher to get out of the ice. A clearance to climb to a higher altitude was issued by Denver Center. Shortly, the pilot reported that he couldn't climb or maintain altitude. A few seconds later he announced that he was spinning in. The aircraft disappeared from radar and a probable crash was reported to the local sheriffs department.

Significant factors with this accident are that the aircraft crashed within 8 miles of a lighted mountain airport where the ceiling was approximately 5,500 feet with good visibility. Could the pilot have asked for vectors to this airport? The report was that **the pilot needed to get to work the next morning.** The pilot received an extensive weather briefing but asked few, if any, questions. The amount of information given the pilot was so much that he seemed not to fully understand how the variety of weather, present and forecast, related to his flight. Careful enroute flight management may well have had this pilot consider diverting to an enroute alternate airport and wait for improved conditions.

CONCLUDING THOUGHTS

CONDUCT ALL IFR OPERATIONS AS A COMPETENT MANAGER/PILOT. POOR DECISIONS PROMOTE UNSAFE OPERATIONS AND ACCIDENTS WHICH COULD HAVE BEEN PREVENTED.

Avoid any attitude or mindset which fails to consider alternate courses of action which may enhance flying safety. IFR operations require an extra measure of professionalism in all pre-departure decisions and operations.

While enroute the pilot must apply sound management principles and judgements. The success and safety of each IFR operation is heavily dependent upon the management efforts of the pilot. Careful compliance with the principles of COCKPIT RESOURCE MANAGEMENT would prevent most IFR accidents. Review Chapter 1.

REVIEW OF DEFINITIONS, ETC.

This chapter is focused on the single engine limitations and capabilities of light twins. Most accidents, unique to light twins, involve improper management and flying of the aircraft after an engine failure or significant engine malfunction. When the light twin is involved in normal operations, the accident causes pretty much parallel the causes associated with the high performance single engine aircraft.

Let us begin by reviewing the following key information which is contained in the pamphlet entitled *"Flying light twins safely"*. This pamphlet is distributed by the FAA and is great study document.

Know the key airspeeds for your airplane and when to use them:

V_{mca}(Red Radial) - airspeed below which directional control cannot be maintained with one engine inoperative.

V_{sse} - never intentionally cut an engine below this airspeed.

V_{yse}(Blue Radial) - always fly this airspeed during a single engine emergency during climb-out (except when necessary to clear an obstacle after take-off) and on final approach until committed for landing.

V_{xse} - Fly V_{xse} to clear obstacles, then accelerate to V_{yse}.

Know the performance limitations of your airplane including its:
- accelerate-stop distances,
- accelerate-go distances,
- single engine service ceiling, and
- maximum weight for which single engine climb is possible.

The single engine service ceiling is the maximum altitude at which an airplane will climb, at a rate of at least 50 feet per minute in smooth air, with one engine shut down. New **Handbooks** show service ceiling as a function of weight, pressure altitude and temperature while the old **Flight Manuals** frequently use density altitude.

The single engine service ceiling chart should be used during flight planning to determine whether the airplane, as loaded, can maintain the **Minimum Enroute Altitude** (MEA) if IFR, or terrain clearance if VFR, following an engine failure.

Climb performance depends on four factors:
- Airspeed - too little or too much will decrease climb performance
- Drag - gear, flaps, cowl flaps, windmilling prop and speed.
- Power - amount available in excess of that needed for level flight.
- Weight - passengers, baggage and fuel load greatly affect climb performance.

CEILING
The ceiling as used here defines the highest elevation at which the aircraft will maintain altitude or the highest elevation to which the aircraft can climb during the stated configuration and conditions and with average pilot skills.

ENGINE FAILURE DURING TAKEOFF

> IF AN ENGINE FAILS BEFORE ATTAINING LIFT-OFF SPEED, THE ONLY PROPER ACTION IS TO DISCONTINUE THE TAKEOFF. SEE FIGURE 13-1.
>
> IF THE ENGINE FAILS AFTER LIFT-OFF, WITH THE LANDING GEAR STILL DOWN, THE TAKEOFF SHOULD BE DISCONTINUED IF TOUCHDOWN AND ROLL-OUT ON THE REMAINING RUNWAY IS STILL POSSIBLE.
>
> IF YOU DO FIND YOURSELF IN A POSITION OF NOT BEING ABLE TO CLIMB, IT'S MUCH BETTER TO REDUCE THE POWER ON THE GOOD ENGINE AND LAND STRAIGHT AHEAD THAN TRY TO FORCE A CLIMB AND LOSE CONTROL.

Keep in mind that the FAA certification regulations **do not** require that a light twin be capable of climbing or maintaining altitude on one engine if the certified maximum gross weight is less than 6,000 lbs. The objective of this chapter is to expand the reader's understanding of the performance and operational limitations and capabilities which apply during single engine operations.

> NOTHING IN THIS CHAPTER IS INTENDED TO ADVOCATE OR ENCOURAGE ANY ACTION OR PROCEDURE WHICH IS IN ANYWAY CONTRARY TO PROCEDURES OR PRACTICES SET OUT IN THE FAA APPROVED POH FOR THE AIRCRAFT BEING OPERATED.

If safety is to be served, the light twin pilot must not expect the aircraft to do something which is beyond its' performance capability. Know what the airplane can do on one engine in all operating situations.

Equally important is for the light twin pilot to not attempt any single engine operation which may be beyond his/her capability. This single engine capability varies among pilots.

Pilot proficiency depends on the quality of the original twin engine training, the total flight hours, and the light twin flight hours. An additional and extremely important factor is the frequency of light twin recurrent training and operating experience.

WHAT IS SIGNIFICANT AFTER AN ENGINE FAILS

When operating on one engine the ratio of gross weight to HP available is doubled. With all the weight but only half the normal power the degradation of performance is very significant. Directional control including rudder trim become an ongoing chore. Single engine cruise airspeed will be reduced to around two-thirds of normal. Single engine climb rate and ceiling will be much reduced. The appropriate S.E. pitch attitude must be maintained throughout.

The following examples will portray the ratio of gross weight to HP available at various elevations.

GROSS WT. VS. HP AVAILABLE - AT SEA LEVEL

Let's assume that we are flying a light twin with a gross weight of 5,300 lbs. powered by two engines of 285 each. The prevailing conditions are O.A.T. 59 degrees F and the elevation is sea level.

In the above normal 2 engine operation the ratio of gross weight to HP available is **9.29 lbs.** per 1 HP.

After an engine is shut down the ratio will be **18.58 lbs.** per 1 HP.

Assuming the same prevailing conditions a Cessna 150 would have a ratio of gross weight to HP available of **16.00 lbs.** per 1 HP. Compare this to the 18.59 lbs. per 1 HP for the single engine operation of the light twin which is being discussed here.

This ratio is derived by noting that the Cessna 150 has a maximum gross weight of 1600 lbs. and is powered by a 100 HP Continental engine.

A brief comparison of the above numbers is:
Light twin - 2 engines at full power = 9.29 lbs. per 1 HP.
Light twin - 1 engine at full power = 18.58 lbs. per 1 HP.
Cessna 150 - 1 engine at full power = 16.00 lbs. per 1 HP.

In these examples the Cessna 150 will have a climb rate of 670 FPM according to the POH. The light twin on one engine can be expected to produce a climb rate ranging from approximately 300 FPM to 450 FPM depending upon the make and model of aircraft, gross weight if the operation is at sea level and O.A.T. is 59 degrees F. At higher density altitudes the single engine performance will be severely degraded.

GROSS WT. VS. HP AVAILABLE - AT S.E. CEILING

The single engine ceiling will be approximately 7,500 feet MSL for many light twins when the gross weight is near maximum and the O.A.T. is at standard temperature.

Keep in mind that a normally aspirated engine suffers a decrease in power output equal to approximately 3 percent for each 1,000 feet above sea level. Humidity and temperature also impact on this percentage of power decrease. The number of 3 percent decrease in power output for each 1,000 feet above sea level is a guideline accurate enough to serve operational needs.

At an elevation of 7,500 feet MSL, which is the S.E. ceiling used during this discussion, a 285 HP engine will be developing only about 215 HP. This is with maximum RPM and full throttle. When operating on one engine with the gross weight at 5,100 lbs. The ratio of gross weight to HP available will be **23.76 lbs.** per 1 HP. We can now see that single engine performance at or near the S.E. ceiling is marginal.

The airplane can climb no higher than the present altitude. Additionally, the airplane cannot maintain altitude at an elevation where the ratio of 23.76 lbs. per 1 HP is exceeded. The light twin is now at its' single engine ceiling.

GROSS WT. VS. HP AVAILABLE - WHEN ABOVE THE S.E. CEILING
For this example we assume that the light twin has a gross weight of 5,000 lbs. and is cruising at 10,000 feet MSL. The O.A.T. is standard at which point the single engine is producing approximately 200 HP at full throttle and maximum RPM. The ratio of gross weight to HP available is now **25.50 lbs.** per 1 HP compared to **23.76 lbs.** per 1 HP at the S.E. ceiling. WHAT WILL HAPPEN HERE?

DRIFT DOWN
Will the airplane be capable of maintaining 10,000 feet MSL on one engine? **The answer is an emphatic NO.** With full power on the good engine, the airplane properly trimmed and the airspeed on the BLUE LINE -- the aircraft will gradually descend to its' single engine ceiling. This is referred to as a **DRIFT DOWN.** See FIGURE 13-2.

THE TERM **DRIFT DOWN** IS USED TO DESCRIBE A SITUATION WHEREIN A MULTI-ENGINE AIRCRAFT, WITH AN ENGINE(S) SHUT DOWN IS UNABLE TO MAINTAIN THE PRESENT ALTITUDE AND IS FORCED TO DESCEND TO A LOWER ALTITUDE WHERE LEVEL FLIGHT CAN BE MAINTAINED WITH THE POWER AVAILABLE. See FIGURE 13-2.

The **DRIFT DOWN** will continue until the airplane has descended to its' single engine ceiling which will depend upon the gross weight and the O.A.T.

As a point of interest a Boeing 727 is also subject to **DRIFT DOWN** as are all multi-engine aircraft.

If a B-727 is cruising at FL-390 when an engine is shut down the subsequent drift down would continue to around FL-260 depending on gross weight, OAT and which model of B-727 is involved. If two engines are shut down the one engine ceiling will be approximately 14,000 feet.

DRIFT DOWN - TYPICAL DESCENT RATES

The drift down rate during single engine operation of a light twin, when 3,000 feet above the S.E. ceiling, will usually begin at about 250-300 FPM. The rate of descent will gradually decrease as the aircraft descends and the operating engine develops approximately 23 inches manifold pressure. At 1,000 feet above the single engine ceiling the rate of descent may be around 100 FPM or less. The descent rate will continue to decrease until it reaches zero at the S.E. ceiling.

DIRECTIONAL CONTROL - DURING S.E. OPERATION

With either engine failed the resulting yaw force will be strong.

Immediate and aggressive application of rudder is required in order to maintain the desired heading. The heading must be maintained, otherwise, as the airplane yaws abruptly toward the dead engine the bank angle can get out of hand. The resulting corrective action will be an unsettling experience. Altitude lost and heading deviations may be significant. All this trouble can be avoided by staying on heading using whatever rudder pressure is necessary. It is for this reason that the heading bug should be continuously set to the heading being flown.

While holding on to the desired heading, immediately adjust the nose up pitch to the pitch attitude appropriate for single engine operation. This pitch attitude must be derived for the airplane being operated. See the takeoff profile chart 13-1 at the end of this chapter. Adjust rudder and elevator trim as soon as possible.

To minimize drag during S.E. operations the airplane should be banked into the dead engine approximately 5 degrees and the rudder trimmed to maintain desired heading. The ball will be about one ball width from center toward the lower wing. Check the applicable POH.

DON'T BE FOOLED BY THE S.E. PERFORMANCE DURING TRAINING

Single engine flight training is customarily conducted with only the trainee pilot and the CFI on board. Additionally, there is usually less than a full fuel load and no baggage. The gross weight is then 400 to 600 lbs. less than the maximum gross weight for the aircraft. A significant fact is that the S.E. ceiling, of the aircraft during training, may be as much as 3,000 feet above the POH number for maximum gross weight. A light twin operating at **training weight** will also climb at a much higher rate of climb due to the lower gross weight.

For *"real world"* operations always assume that aircraft performance will barely equal the POH numbers for the aircraft at maximum gross weight and operating on one engine.

POINTS TO PONDER

When operating on one engine, the cruise airspeed will decrease by approximately one-third. Typically, a two-engine cruise airspeed of 160 kts. will decrease to about 110-120 kts. after an engine is shut down. The point is that a specified distance flown on one engine will require about as much fuel as if flown on two engines.

At full throttle and maximum RPM the good engine fuel consumption will increase significantly below about 6,000 feet MSL because the engine will be producing more power due to higher manifold pressure.

For example, during two engine cruise each engine may burn 12 gph. The same light twin operating on a single engine at full power may burn 18 gph when around 3,000 feet MSL or lower and travelling at a reduced airspeed.

The point to be understood here is that because of the lower airspeed during single engine flight and with the higher per hour fuel consumption on the good engine -- a S.E. flight of 75 to 80 miles will require as much or slightly more fuel than the same distance in normal 2 engine cruise. Here is the problem and the answers. Let's look at a typical situation. First, during two engine cruise at 160 kts. for a distance of 80 NM and with fuel consumption of 12 gph per engine. The flight time would be 30 minutes and total fuel consumed would be 12 gallons.

In a single engine operation at or below the S.E. ceiling, at full power and IAS of 110 kts. the same 80 mile flight would require about 44 minutes and at a fuel flow of 18 gph. The amount of fuel consumed would be about 11 to 13 gallons depending upon how low the S.E. flight was conducted. This compares to 12 gallons in the two engine situation just discussed.

This example is intended to highlight two points. First, a single engine operation at lower altitude will burn about the same amount of fuel per mile as a two engine operation. Second, do not fail to promptly check the fuel quantity available to the good engine after an engine shutdown. Anytime a single engine operation will not put the airplane on an airport in 15 minutes or less it is necessary that fuel cross-feed operation be considered and set up if advisable to do so. Do **not** run out of fuel on the good engine! When this has happened the **result** has usually been catastrophic! Always use the POH procedure to set up a cross-feed to the good engine.

CHECK LIST REQUIREMENTS FOR SAFE OPERATIONS

It is my opinion based on more than 50 years experience as a general aviation CFI, as a charter pilot and as an airline pilot and instructor, that hand held or similar printed check lists are an absolute necessity for safe and professional quality operation. I encourage each light twin pilot and CFI to make any additions to the POH check lists which would enhance the completeness of the check list and thus enhance the safety of operations.

A suggested revision is to duplicate the **ENGINE FAILURE ON TAKEOFF** check list and display this extra list immediately following the run up or before takeoff check list.

This key check list will then be conveniently located so that it can be reviewed **prior to each and every takeoff.** This review should be done without exception!

By following the suggested sequence the engine failure on takeoff will be fresh in mind each time a takeoff is begun. A pilot may fly the same aircraft for 25 years and not experience an engine failure on takeoff or initial climb. Such a pilot is a very lucky aviator. Do not bet your life and the lives of your passengers that an engine failure will not occur during this critical segment of flight. If an engine failure does occur at a critical time the extra review will probably save the airplane and the occupants. This is a great return on a small investment of time.

ENGINE START AND TAXIING

These operations in a light twin are so straight forward that discussion here would be of little or no benefit.

THE RUN UP

Complete the run up, including the prop feathering check, in accordance with the check list. The prop feathering check is of the utmost importance and must comply with the check list in all functions -- otherwise it is **NO GO.**

IF EITHER PROP FAILS THE FEATHERING CHECK THE AIRCRAFT MUST NOT BE FLOWN UNTIL REPAIRS HAVE BEEN MADE AND THE FEATHERING CHECK IS SATISFACTORY.

If, while airborne, the prop will not feather, the penalty in performance will be alarming! The only action to help this situation is to position the prop control full aft.

This will move the prop about one-third of the travel toward the full feathered position. Prop drag will be significantly reduced but the ability to maintain altitude will be doubtful. If there is no control of prop pitch or the feathering function, the pilot is then looking at a very serious problem. The best solutions may be to proceed to a nearby airport or consider an off airport landing before running out of altitude.

There are many makes and models of props and feathering systems. A light twin pilot must have a thorough understanding of the prop system for the aircraft being operated.

There are several common causes for a prop to fail to feather. The malfunctions may involve cables, pulleys, fittings, black boxes, valves or seals.

THE TAKEOFF

Make a power check as soon as power is applied to begin the takeoff. RPM, manifold pressure and fuel flow/mixture must each be in the normal range. A second power check should be made at 10 to 15 kts. prior to the beginning of the rotation for lift off. Keep the right hand on the throttles throughout the takeoff roll and initial climb to 1,000 feet or more above the airport. The right hand should be removed from the throttles only to operate the gear switch or other essential controls.

The proper pilot action in event of engine failure or malfunction, prior to reaching the V_{mc} speed, is to instantly close both throttles! If after a power failure below V_{mc}, the asymmetrical power is left applied the airplane will be pulled off the runway in a heartbeat. This loss of directional control cannot be stopped by the use of brakes and rudder control. The only remedy is to close both throttles immediately!

The loss of directional control, with an engine failure, will be very abrupt if the airspeed is low - say 40 kts. At this speed there is very little rudder authority. By contrast, an engine failure, with the airspeed slightly less than V_{mc}, will result in a more gradual loss of directional control -- but a loss of control nonetheless.

Some procedures may call for lift off prior to reaching V_{mca} airspeed. Should an engine failure occur with the aircraft airborne below V_{mca}, directional control will be lost and the wing on the dead engine side will drop abruptly -- possibly contacting the ground. The only safe action in this situation is to close the throttles. There will then be sufficient rudder authority to maintain directional control.

The safest practice, in my opinion, is to begin rotation just prior to reaching V_{xse} airspeed. With this practice the main wheels will lift off at about V_{xse}. The acceleration in ground effect will be rapid with V_{yse} and V_y speeds being reached in a few seconds. See FIGURE 13-1.

Keep in mind that manufacturers usually specify a lift off speed which is lower that V_{xse}. Check you POH. This lift off speed must be adhered to if the takeoff distance chart numbers are to be valid. If you do lift off approaching V_{xse} as I have mentioned above, an additional takeoff distance, and distance to clear a 50 ft. obstacle, will be required. It is my experience that adding about 15 to 20 percent to the POH chart values is adequate.

In day to day operations utilizing runways far longer than needed then the extra 150 to 200 ft. of ground roll to near V_{xse} makes for a cleaner and safer lift off -- especially in cross winds, or with a heavily loaded aircraft, at night and for takeoffs in high altitude and high density altitude situations.

If there is **any doubt** about required runway length or if you fly a twin which is subject to WHEELBARROWING at higher takeoff speeds -- then by all means use the POH lift off speeds. This is a judgement call for each P.I.C. to make. My opinion on light twin lift off speeds is presented here as additional information which the reader may find useful.

Make a practice of rotating to a predetermined nose up pitch attitude during every takeoff. For several light twins 10 degrees nose up pitch works very well.

INITIAL CLIMB

The initial climb should be made at V_y airspeed and with full takeoff power to 1,000 feet or higher above the airport. If an engine failure occurs during this flight segment, lower the pitch attitude to the single engine nose up pitch attitude which will produce **BLUE LINE** airspeed. Maintain directional control by promptly applying rudder toward the good engine. THINK: *"Keep the wings level with rudder"*! Secure the failed engine. The drill is to maintain the heading with rudder and set the nose up pitch with the elevator. Next declare an emergency on whatever frequency is being used for communications. Proceed to the nearest suitable airport and land. In most instances the best choice is to return to the departure airport.

Allow sufficient time to communicate and plan the approach. Complete all applicable check lists and keep airspeed under control. A most serious problem during single engine approaches is to have excessive airspeed. The single engine approach will usually be flown with gear down and flaps at zero. With this low drag configuration and the airplane in a descent, it is very easy to be far too fast. Proceed with a methodical and unhurried operation. Haste, poorly planned and poorly monitored operations have transformed many simple single engine approaches into catastrophic events. **Slow down and think.** See FIGURE 13-3.

If you have the bad luck of an engine failure during takeoff or initial climb from an airport with an elevation higher than the S.E. ceiling of the aircraft being operated your options are very limited.

If the altitude at engine failure is around 1,000 feet above the airport or higher then it may be possible to get back to the runway safely while working with a **drift down** rate of 200 FPM or so. If the power failure occurs at a low altitude on departure, from a high altitude airport, then the best plan may be to look for an off airport landing site and make a single engine approach to it. Here again we can see the wisdom of departing from high elevation airports at well below maximum takeoff gross weight. This would increase the S.E. ceiling.

CRUISE CLIMB, CRUISE AND DESCENT

The key factor here is whether or not the aircraft can maintain the current altitude in case of an engine failure. If an engine failure occurs while above the S.E. ceiling, the pilot will be confronted with **drift down** which was previously discussed.

Keep in mind that while most light twins will have a single engine ceiling of roughly 7,500 feet MSL there are some lower powered twins with a S.E. ceiling around 4,000 feet to 5,000 feet MSL. Be sure to consult the POH for the aircraft being operated.

THE SINGLE ENGINE APPROACH AND LANDING

A single engine approach and landing is a very demanding task unless you have had recent refresher training and practice. The single engine approach and landing presents an unfamiliar situation thus it is extremely important for the pilot to remember **the differences** which apply to the single engine versus the 2 engine approach and landing.

FIRST - single engine approaches are flown in a low drag configuration. Airspeed control is difficult. Most S.E. approaches will be flown with gear down and flaps at zero. Keep in mind the time worn phrase -- *"when the gear goes down the airplane goes down"*.

The gear may be extended anytime a continuous descent can be made to touchdown. The gear must be extended early enough to allow time to stabilize the airspeed and rate of descent -- and complete the landing check list. I recommend that you refer to the previously discussed **PRECISION VISUAL APPROACH.** The single engine approach profile should be at a normal descent rate. The reason being that the S.E. approach profile will then appear as normally seen by the pilot. A single engine approach is not the place to engage in either a high or flat approach. The pilot will have a hand full flying the single engine approach utilizing the familiar final approach profile.

SECOND - Airspeed control is difficult and requires practice. The best possible single engine approach will result when the pilot knows a basic power setting for the final approach. Check the power setting chart at the end of this chapter. Note the sample configuration and power chart, FIGURE 13-3. **Fly your own airplane** to gather the numbers to fill in the blanks. A reliable M.E. CFI should be on board during the gathering of these numbers.

THIRD - Elevator and rudder trim requirements are strange and demanding as compared to operating a normal approach and landing. The elevator trim operation will be straight forward except in a situation where flaps are extended from zero to one-half or full flaps when you *"have the runway made"*. This abrupt flap extension will cause a significant airspeed decrease and may upset the pitch trim at a very awkward time during the approach. If flaps are extended here be prepared to hang on to the pitch attitude by aggressive elevator pressure - either nose down or nose up as required. There may not be enough time to retrim at this critical moment and thus it may be safer to *"strong arm"* the elevator until the aircraft is landed.

Rudder trim is necessary during a single engine approach and landing. This trim activity is ongoing and very distracting unless steps are taken to minimize the need for rudder trim adjustments. This brings us back to the need to have predetermined power settings throughout the single engine approach operation. The key is that each change in power setting and/or airspeed requires a rudder trim adjustment.

This need for rudder trim is extremely important during the landing flare and the touchdown. **During the S.E. approach the rudder will be trimmed toward the good engine.**

As power is reduced the trim will attempt to turn the aircraft in the direction of the good engine. Example - the S.E. approach is being flown with only the right engine operating. There will be a goodly amount of right rudder trim in place to keep the right engine from turning the aircraft to the left. But, as we begin to reduce power for landing, the rudder trim, which is deflecting right rudder, will begin to turn the aircraft to the right. You should expect this turning tendency to be very strong. Appropriate trim and/or rudder adjustments must be made promptly.

There are various ways to maintain directional control with the rudder **during the flare and touchdown.** One method is to use strong and aggressive rudder applications against the trim. This method may appear to be a bit awkward, but it does avoid the distraction of operating the rudder trim at a moment when you need more hands. Another method is to remove at least part of the rudder trim during the power reduction while beginning the flare. Another method is to brief the right seat passenger to adjust the trim by your instructions i.e. *"start now"*, *"stop"*, *"a bit more"*, etc. while you reduce power during the landing flare.

In this way any additional *"leg"* trim will be lessened and easy to accomplish. The reliability of the co-pilot must be carefully considered before utilizing this method of trim adjustment.

When is the rudder adequately trimmed during a single engine approach? My answer is at variance with the opinions of some CFI's. I am a firm believer in having the aircraft trimmed as well as is practical. However, I am willing to deviate slightly during a single engine approach flown by a single pilot.

The pilot who puts too much attention to micro adjustments of rudder trim during final approach runs the risk of the approach getting off profile and the airspeed getting out of hand. This can easily result in an unsafe approach. Let me reiterate that I am not saying disregard the rudder trim during final approach in a light twin operated by a single pilot. What I am saying is that rudder trim will be adequate if the airplane goes straight with a bit of rudder pressure applied occasionally. Be very aggressive in maintaining alignment with the runway centerline.

AUTO PILOTS AND RUDDER DISCIPLINE

A light twin pilot must be continuously prepared to make positive rudder inputs. Even when operating with the auto pilot engaged an engine failure will cause such an abrupt directional swing that the auto pilot will probably be kicked off by the excessive turning moment. The result will be a wild ride which can see the airplane rolled against the dead engine possibly even to the inverted attitude. You may remember a China airline's B-747 which was over the Pacific Ocean about 600 miles from San Fransisco when an outboard engine ran out of fuel. The airplane swerved toward the dead engine and the auto-pilot disengaged. The airplane rolled nearly inverted and went into a screaming dive toward the ocean. As the airplane descended into the *"real atmosphere"* the aircraft began to seek its trimmed IAS and recovered at a low altitude. A section of the stabilizer and elevator separated from the airplane. The crew was successful in crippling into San Fransisco and landing safely. This whole episode may have been prevented had either the captain or co-pilot been alert at the controls including on the rudder pedals. So there you have it -- a long recitation regarding what I think is important -- **keep those feet on the rudder pedals at all times auto-pilot or not!**

THE GO-AROUND OR MISSED APPROACH

A go-around or missed approach during single engine operation can be accomplished as long as all the operation is being conducted well below the S.E. ceiling of the aircraft being operated. Be alert to a sudden and strong turning moment toward the dead engine as maximum power is applied to the good engine. Rotate to the pre-determined nose up pitch attitude for V_{yse}. This is a major key to airspeed control. Begin flap and gear retraction, per the POH, then adjust rudder and elevator trim.

Here the objective is to make a smooth and positive transition from a single engine approach to a single engine initial climb. If the airplane is other than very lightly loaded you should expect climb performance to be very poor. **Patience** is the key word here. The poor performance on this single engine maneuver should be taken as powerful motivation to avoid a single engine go-around or missed approach.

THE CARE AND FEEDING OF A SICK ENGINE

If an engine begins to run rough, backfire or surge, at a medium to high power setting, you may be able to coax the engine to continue to run but at a reduced power setting. Be sure to check oil pressure and temperature, CHT, fuel flow and mixture setting.

The following discussions of engine problems is not intended to lead any pilot to deviate from any POH procedure or any operating recommendations which may be set out in the POH. The purpose is to highlight common symptoms of engine problems and to state what action has been safe and satisfactory for me and many other pilots.

STRUCTURAL FAILURE WITHIN THE ENGINE

If the engine is pounding and vibrating severely you may have a heavily damaged engine. Shut it down and proceed to a suitable airport. These situations sometimes spawn engine compartment fires. Be alert for any indications of fire.

FUEL FLOW PROBLEMS

Fuel flow problems usually involve fuel exhaustion in the tank selected. There are other problems lurking and they are reviewed here. An engine driven fuel pump may fail completely or partially. The remedy is to turn on the boost pump in accordance with the POH procedure.

Fuel flow may be restricted to such an extent that only a fraction of the normal fuel flow volume gets to the engine. The remedy is to reduce the throttle setting and perhaps the engine will then run at partial power. Another common cause of low fuel flow is a fuel tank selector which is not precisely in the detent. Then there is improper use of the mixture control. This situation appears during descent from higher cruise altitudes where the mixture has been leaned significantly. Subsequently, the descent is made and the mixture is not enriched appropriately. If power is applied at lower altitudes for a go-around, missed approach or other operation a power interruption may occur. This situation is the equivalent of leaning the mixture until the engine dies while at a high power setting. The obvious remedy is to enrich the mixture until the engine runs normally. Most fuel flow problems are best addressed initially by following the time honored POH/FAA statement which says switch the fuel selector to another tank and adjust the position of the mixture control as appropriate.

HAVING ONE ENGINE AT NORMAL POWER AND THE OTHER ENGINE AT PARTIAL POWER IS MUCH SAFER THAN A SINGLE ENGINE OPERATION. THE LIGHT TWIN PILOT SHOULD HANG ON TO ANY ENGINE WHICH CAN BE KEPT RUNNING AND DELIVERING POWER -- PROVIDED PRESSURES AND TEMPERATURES ARE IN A SATISFACTORY RANGE.

When making long climbs at high power settings and high O.A.T. a fuel flow interruption may occur. This is caused by excessive heat in the engine compartment resulting in vapor lock. The cure for this is to be sure the cowl flaps are full open, turn on the aux fuel pump, lower the nose and accelerate while reducing power if practicable to do so. The objective is to increase the air flow through the engine compartment so as to reduce the air temperature and cool the fuel lines. Check the POH for the airplane being operated and follow the applicable procedure. If the POH specifies cowl flaps open during climb -- comply!

IGNITION PROBLEMS

Backfiring and roughness are associated with ignition problems. These problems may be caused by a spark plug which is misfiring, not firing or by a malfunctioning magneto. The corrective action is to reduce manifold pressure to 18 to 20 inches and enrich the mixture a bit. If engine operation does not smooth out, then follow on with a magneto check. This check will indicate whether the problem is with either the left or the right ignition system and whether or not the engine operates satisfactorily on only one ignition system.

If the engine runs smoothly on one magneto then proceed at reduced manifold pressure of about 20-22 inches. Readjust the mixture to a slightly richer setting and proceed to a suitable airport.

With only one spark plug firing in a cylinder, the fuel burning pattern within the cylinder is not normal. Reduced manifold pressure and a cooler mixture will help prevent damage due to uneven heat build up and dispersal during each firing cycle. It then goes without saying that a takeoff should not be attempted while operating on only one ignition system.

FUEL CONTAMINATION

Fuel contamination is a deadly problem! Fuel contamination takes many forms. The list includes water, rust, bits of debris, dirt, sand and jet fuel. These things get into fuel tanks through fuel caps that do not seal properly, fueling errors, etc. This contamination must be prevented by all means possible. Fuel contamination accidents usually occur at low altitude and are frequently fatal to all on board.

SUGGESTED RECURRENT S.E. PRACTICE

Do this practice with the most experienced multi-engine CFI you can find.

PRE-FLIGHT BRIEFING - 1:00

Review of S.E. procedures, plus S.E. operational limitations and capabilities.
Review of the profiles for S.E. initial climb, S.E. approach and landing, plus go-around/missed approach.
Complete briefing for the practice period.

PRE-FLIGHT INSPECTION - 00:30

This pre-flight should be thorough and include any review of discrepancies discovered during previous inspections.

FLIGHT PERIOD - 1:30

First takeoff - Have a simulated engine failure when the airspeed is 15 kts, less than the rotate speed. The instant anything abnormal is detected close both the throttles immediately and concentrate on directional control and stopping. Return to two engine operation for the taxiing for the next takeoff.

Second takeoff - Make a normal lift off and at about 300 feet AGL a simulated engine failure will be introduced. Remain on the desired heading by prompt application of rudder while establishing the nose up pitch attitude which is appropriate to maintain **BLUE LINE** airspeed with one engine operating. Simulate the steps in the engine shut down procedure. The CFI will have set a zero thrust power setting to simulate the power failure. Simulate declaring an emergency. Continue S.E. climb to 1,000 feet AGL. Return to a normal two engine operation. Establish the enroute climb profile. Continue climb to 3,000 feet AGL. Maintain 3,000 feet AGL.

The next activity is to practice an actual engine shut down at the moment of a simulated lift off. This is done at 3,000 feet AGL. The exercise is accomplished by configuring the aircraft for the takeoff roll and the lift off. When the aircraft has been slowed to rotate airspeed, takeoff power is applied and rotation is made to the two engine nose up pitch attitude appropriate for a normal takeoff. As soon as the pitch attitude is established the **CFI will announce** *"ENGINE FAILURE"* while moving a throttle smoothly to idle.

Accomplish the engine failure on takeoff procedure per the POH. Continue the S.E. climb for 500 feet or more while noting the S.E. rate of climb -- if any!

Next, discontinue the climb and remain on **BLUE LINE** airspeed while the CFI adjusts the good engine power to simulate **DRIFT DOWN** of 200 FPM. Continue with the drift down until reaching 2,000 feet AGL. Next proceed to the airport for a single engine approach and landing. During the S.E. flare and landing be alert to the need to control heading by the timely application of rudder and/or rudder trim to compensate for removal of the asymmetrical thrust.

POST FLIGHT CRITIQUE - :45

This period should be a discussion and review of all operations carried out in the recurrent S.E. practice.

COULD THESE ACCIDENTS HAVE BEEN PREVENTED?

A twin engine aircraft was on takeoff roll at about 50 kts. airspeed when an engine failed abruptly. The thrust of the good engine promptly turned the aircraft off the runway at a sharp angle. There was a delay in closing the good engine throttle. The airplane came to rest a good distance from the runway edge, and suffered substantial damage. The pilot was uninjured.

A light twin was returning from a night cross country when it crashed more than a mile short of the runway. The pilot had been cleared for the approach and subsequently reported over the outer marker inbound. When the wreckage was located both fuel selectors were positioned to off. There were no survivors.

A twin was making an early morning arrival at a busy airport in CAVU weather conditions. The trip was for business purposes and the occupants were the owner and some employees. While on the downwind leg one engine died when the selected tank ran dry.

During the ensuing confusion and/or inappropriate action the airplane went out of control and crashed. All occupants were killed.

A light twin was being flown by the owner to another airport for an annual inspection. Shortly after takeoff the pilot reported an uncontrollable engine fire. The airplane crashed shortly thereafter and the pilot was killed. During the investigation the engine exhaust stack was found on the ramp where the engines were started. The fire was caused by the red hot exhaust being pumped into the engine compartment until a fire was started.

Takeoff Profile for Twin Engine Aircraft

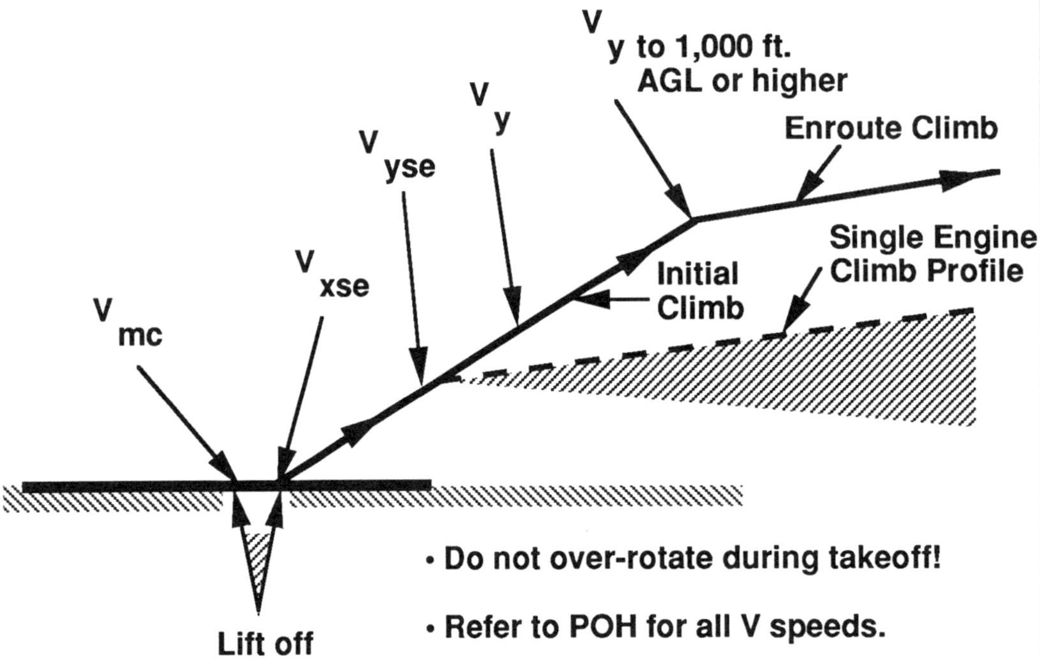

- Do not over-rotate during takeoff!
- Refer to POH for all V speeds.
- — — — Portrays S.E. climb out at Vyse (Blue Line IAS) after a power failure - if S.E. climb capability exists.

- Begin rotation so that lift off will occur at Vxse.
- Rotate to the pitch attitude for climb at Vy (Note this pitch attitude for the twin you fly).
- Vyse will usually be attained at about 50 ft. AGL.
- Note the normal rate of climb during each initial climb. If the climb rate is significantly below normal, check power settings, gear, flaps, etc.

Figure 13-1

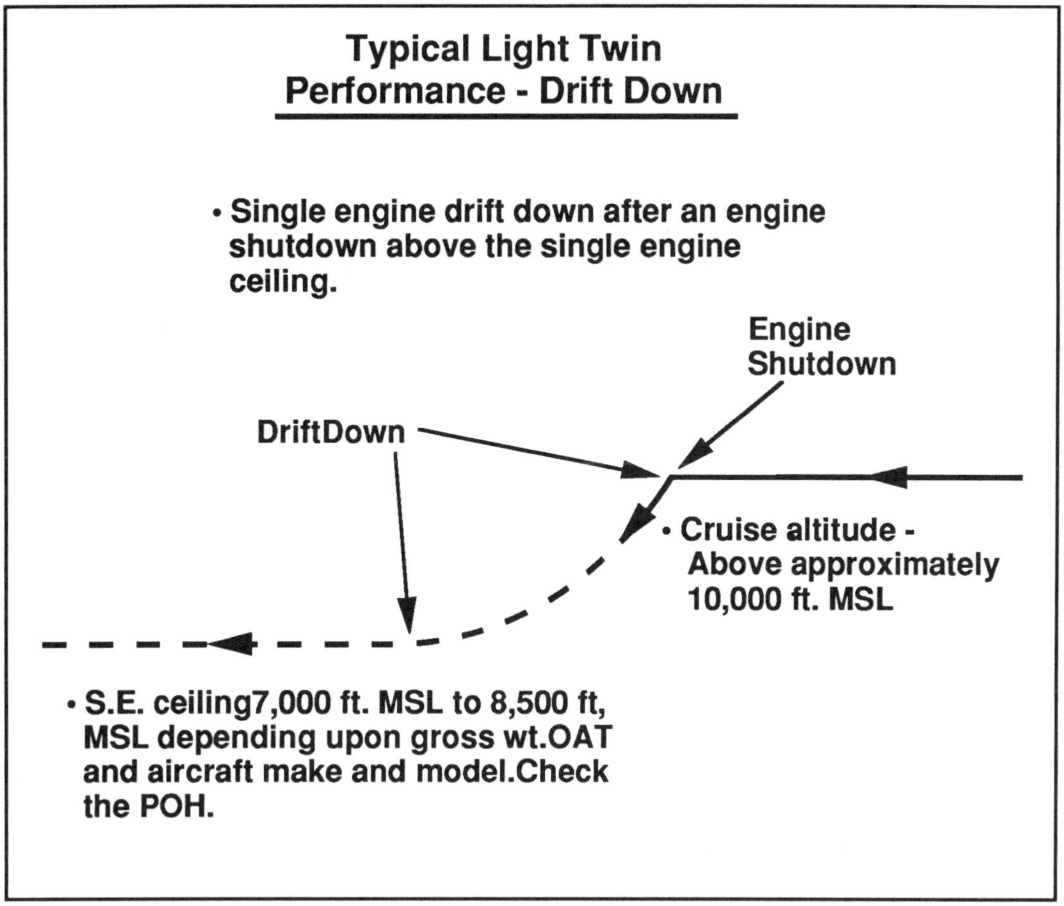

Typical Single engine drift down/climb
(Temperature - standard + 40° F.)

Engine failure altitude	5,100 lbs.	4,500 lbs.
12,000 ft.	Descend 250 fpm	May hold altitude
9,000 ft.	Descend 100 fpm	Climb 75 fpm
7,000 ft.	May hold altitude	Climb 150 fpm

Figure 13-2

THE S. E. APPROACH - - IN A LIGHT TWIN

CONFIGURATION = Gear __Down__

Flaps ________

POWER = RPM __Max.____

MP ____________ Inches

AIRSPEED = __Blue Line IAS__ (Until landing is assured)

RATE OF DESCENT = ____________ FPM (Approx.)

AIM POINT = ___________ Ft. From Threshold

PLANNED TOUCHDOWN = ____________ Ft. From Threshold

Notes:

1. The landing roll after touchdown should be expected to require approximately 1.5 times the normal distance.

2. The most common error during single engine approaches is to be too fast because the approach configuration will in most instances involve less drag than a normal approach.

3. Consult the POH.

Figure 13-3

THE EVOLUTION OF TAILWHEEL AIRCRAFT

Beginning about 1914 virtually all airplanes were biplanes with tailskids. Tailwheels and aircraft brakes were not yet in use. The tailskids would dig into the turf during all taxiing operations, during the beginning of the takeoff roll and the ground roll after landing. Paved runways had not yet become available even at the busiest civilian and military airports. The airports were mostly laid out so that all takeoffs and landings could be made into the wind. Directional control during takeoffs and landings was manageable largely due to these operations being into the wind and supplemented by the drag of the tailskid digging into the turf. Slight changes of direction were accomplished by abrupt bursts of power which provided sufficient rudder authority to move the tail slightly left of right.

MAKING TURNS IN RAMP OR HANGER AREAS WAS ACCOMPLISHED BY A GAS MAN OR MECHANIC PUSHING AGAINST THE END OF THE WING, WHICH WAS ON THE INSIDE OF THE TURN, WHILE THE PILOT "GUNNED" THE ENGINE. THIS COORDINATED MANEUVER USUALLY BLEW UP A HUGE CLOUD OF DUST BEHIND THE PLANE AS IT SWUNG THRU THE TURN.

Prior to WWII most aircraft were in the tailwheel configuration. This included aircraft operated by military, airline and general aviation pilots. After WWII tailwheel airplanes continued to be built for the general aviation market. Popular taildraggers of the late 1940's and early 1950's included Cessna, Piper, Taylorcraft, Luscombe, Aeronca and Swift.

By the mid 1970's there was an upsurge in T.W. aircraft as war surplus aircraft were restored to flying condition by the civilian owners and various warbird organizations.

These surplus T.W. aircraft included the Stearman, Fairchild PT-19, Ryan PT-22, T-6, BT-13, B-17, C-47, C-45 and others.

Civilian T.W. aircraft now included ag planes, homebuilts, aerobatic planes and continued production of Piper Super Cubs, Citabrias, Cessna 180/185, Cessna 190/195, and many pre-war cabin types, which continued to be operated.

By the early 1980's there was a wave of newer and younger pilots desiring to check out in T.W. airplanes. Most T.W. flight instructors at that time were highly experienced WWII types. As the years have worn on these old timers have decreased in number and we now find that many T.W. instructors are of the younger generation.

TODAY'S WORLD OF T.W. PILOTS

An increasing number of pilots are intrigued by tailwheel aircraft and have a need to know how to operate them safely. These pilots run the full spectrum including private pilots, ag pilots, a variety of bush and charter pilots plus a fair number of airline pilots.

OPERATING T.W. AIRCRAFT TODAY

Aircraft brakes, steerable tailwheels and a few tailwheel locks are now the norm. This transition began in the early 1930's as brakes and tailwheels were gradually introduced. Most takeoffs and landings are now made on paved runways. Gone are the turf or dirt landing surfaces which would absorb takeoff and landing inaccuracies. The gear tires stick firmly to paved runway surfaces thus making correct and accurate takeoff and landing techniques highly desirable.

Listen up and I will outline how to fly a taildragger in a manner which is logical, safe and easy. Be sure to seek out a CFI with a good amount of experience in T.W. flying and instructing. The material in this chapter is intended to provide a broad understanding which is needed if you are to be a proficient T.W. pilot. **No one should attempt to check themselves out in a T.W. aircraft!** You need to have appropriate training by a good T.W. flight instructor.

MY T.W. EXPERIENCE AS A PILOT AND CFI

At the time of this writing my T.W. experience totals approximately 5,000 hours. This experience includes U.S. Army Air Corps flight instructor, during WWII, with approximately 950 hours in the PT-19 aircraft. Later some 1,590 hours of DC-3 time. Other T.W. time is Cessna 195 for 585 hours as charter pilot and Cessna 180/185 for about 250 hours. Additional experience includes the Lockheed 10-A (shades of Amelia Earhart) with some 285 hours as single pilot in charter operations. Then follows a wide variety of T.W. aircraft including Piper, Taylorcraft, Cessna, Luscombe, Aeronca, Monocoupe D-145, Fairchild 24, Stearman, Ryan PT-22, Swift, Rearwin Sportster, Porterfield, Waco UPF-7, Citabria, Beech Staggerwing, Twin Beech, Cessna 195 and Gullwing Stinson. I have instructed in all the listed T.W. aircraft except the DC-3, Fairchild 24, Rearwin Sportster and Waco UPF-7. I had a forced landing while taking off in the Monocoupe D-145 during a flight in the Chicago area in 1947 and a forced landing west of Kansas City in the Cessna 195 in 1952. Both involved engine failures and neither airplane suffered even a scratch.

The listing of T.W. experience is set out to establish my credibility to advocate how to operate a T.W. aircraft during takeoffs and landings and to explain the reasons for these procedures and techniques. During this long period of T.W. experience, I have noted the need to fine tune some of the techniques and procedures which I was originally taught. Thus I have learned, through my T.W. instructing activities, what works best in most situations. Additionally these techniques provide safety and the logic which makes learning and retention relatively easy.

ENGINE START AND TAXIING

During engine starts extra precautions are required in order to avoid injury to persons who may be unseen by the pilot. When starting a radial engine it is important to have a person stand safety watch. This individual should remain in view of the pilot until taxiing is begun. Hand cranking should be done only by a person who is very experienced.

While taxiing, the flight controls should be operated as set out in all the many training documents. Additionally, make a habit of applying full aft control wheel or stick any time power is added during any turn in ground operations -- and throughout the engine runup regardless of the wind direction. Stand on the ramp and take a side look at the angle of the horizontal stabalizer in the 3 point position. It is now easy to visualize that a prop blast with the elevator even in neutral will result in a strong lifting force on the tail.

The taxi speed should be rather slow and with continuous *"S"* turns so that the area ahead of the aircraft can be scanned by the pilot. The Piper Super Cub and the Citibria are examples, of aircraft with flat engines, which offer fair forward visibility.

Single engine airplanes with radial engines allow very limited forward visibility. Examples are the Stearman, Cessna 195, Staggerwing Beech and T-6.

ALL TAILWHEELS ARE NOT ALIKE

The steerable tailwheel is the best type in my opinion. In this system the tailwheel is steered directly by rudder pedal movement via a chain and spring which connect to the tailwheel steering arm. A full swivel feature, of this system, allows the tailwheel to turn freely if a significant side load is applied to the tailwheel. The steerable feature can be kicked out by a brake being applied firmly in the direction of the desired sharp turn. The steerable feature can also be disengaged from outside the cockpit by applying significant side pressure at the leading edge of the horizontal stabilizer. This full swivel feature is utilized to turn the airplane around and push it tail first into a hanger or tie down spot.

A full swivel tailwheel is non-steerable and as the name implies this tailwheel is free to turn 360 degrees in either direction. Fortunately several full swivel systems have a tailwheel lock which allows the pilot to lock the tailwheel in the straight ahead position by use of a cockpit control. Examples of full swivel T.W. with T.W. lock are the Stearman and DC-3. There are, however, several aircraft - mostly older machines with full swivel T.W. systems which do not have a T.W. lock. Pilots of these machines are deserving of the highest respect!

Steerable tailwheel systems and full swivel systems with and without tailwheel locks account for most installations likely to be encountered.

There are, however, some hybrid systems. It is extremely important for the T.W. pilot to understand the operational capabilities and limitations of the tailwheel system at hand.

THE RUN UP AND TAKEOFF

During run up the stick or control wheel must be held full aft in order to prevent the aircraft from being nosed over with the resultant prop strike. When in takeoff position allow the aircraft to roll straight ahead momentarily in order to assure that the steerable feature is engaged in a steerable installation. If the system is full swivel with a T.W. lock this is the time to engage the lock.

Apply back pressure on the stick or control wheel as power is applied to begin the takeoff roll. This **up elevator** application will put a load on the tailwheel so that it will remain firmly pressed against the runway surface as the takeoff roll is begun. The tailwheel is a main source of steering control until the airspeed increases so that rudder authority is adequate. As the takeoff roll is begun rudder authority is minimal while *"P"* factor and any crosswind component are major factors. Thus we very much need the help of the tailwheel for directional control in the early part of the takeoff roll.

STEERING A TAILWHEEL AIRCRAFT, ON THE GROUND, CAN BE ACCOMPLISHED BY THE TAILWHEEL, THE RUDDER, THE BRAKES OR ANY COMBINATION OF THESE. USING THE BRAKES TO STEER SHOULD BE AVOIDED AS MUCH AS POSSIBLE.

After a short ground roll the elevator back pressure should be relaxed smoothly. The tail will now rise unassisted in many cases.

In other situations forward elevator pressure may be required to raise the tail to takeoff position. The appropriate application of elevator pressure to raise the tail varies with the make and model of aircraft, the elevator trim setting and the aircraft C.G..

During takeoff in light wind conditions the ground roll with the tailwheel firmly on the runway may vary from 200 to 400 feet for a Citabria and 400 to 800 feet for a T-6 or a Cessna 195.

How long should the takeoff roll progress before the tailwheel is allowed to lift off the runway in order to assure adequate directional control by rudder alone? The guideline which puts this in perspective is to say that the tailwheel should remain on the runway until the airspeed is at approximately one third of the stall speed. Using this guideline a Cessna 180-H which has a stall speed of 65 MPH would have the tailwheel remaining on the ground until the airspeed is approximately 20 MPH. This must be estimated. The guideline is by no means definite and all inclusive because conditions, aircraft performance and pilot preferences may dictate some variations. Raising the tail too early is undesirable.

The reasons for keeping the tailwheel firmly on the runway at the beginning of the takeoff roll are now readily apparent. The reason for the old open the throttle and get the tail up quickly is not applicable unless the aircraft is equipped with a tailskid! The words **always** and **never** should be used sparingly when applied to flying. Yes, there are exceptions which apply to the question of when to raise the tailwheel during a ground roll for takeoff.

First there are many special situations where raising the tail early is necessary and for a variety of reasons. These reasons include bush operations, test flights, situations where forward visibility is seriously impaired until the tail is raised etc.. It is fortunate that most of these operations involve skilled and high time T.W. pilots.

Second the tail should be raised sooner if the takeoff roll is into a brisk headwind. The speed which applies is **airspeed.** With a little practice the pilot will be able to estimate the appropriate time to raise the tail in a variety of wind conditions.

The following check list format is utilized to set out the procedure for a smooth and well controlled normal T.W. takeoff.

- Align the aircraft on the runway.
- Roll forward a few inches to assure that the tailwheel system is set for takeoff.
- Simultaneously apply takeoff power and up elevator pressure (control wheel or stick aft).
- At the appropriate airspeed relax the elevator back pressure and or assist the tail to rise until a shallow climb pitch attitude is attained.
- Maintain the shallow climb pitch attitude.
- As the **airspeed increases** through and beyond stall speed the aircraft will lift off and blend into a normal climb.
- Maintain directional control throughout with positive applications of rudder.
- In most well executed takeoffs the tailwheel will be only 6 to 8 inches above the runway during the ground roll until lift off. This height will vary depending upon the size of the aircraft. For a DC-3 the tailwheel height at liftoff will be approximately 5 feet.

THE CROSSWIND TAKEOFF

For takeoffs in crosswind conditions we need to avail ourselves of the supplemental directional control afforded by having the downwind aileron deflected to the down position. The down aileron produces much more drag than the up aileron. This is more pronounced at the beginning of the takeoff roll where the airplane is in the 3 point attitude. Since the down aileron is a good distance outboard of the pivot point of the aircraft, this significant leverage will multiply the effectivity of the down aileron.

For crosswind takeoffs we need the directional control afforded by the down aileron along with the control of the tailwheel and rudder.

The name of the game is aileron into the wind on the takeoff roll. Another reason for aileron deflection is to keep the upwind wing from being raised by the crosswind. The factor of directional control is very important during the early part of the crosswind takeoff roll.

How much of the aileron travel should be applied during the takeoff roll? This is a judgement call. The answer is that for a fairly strong crosswind full aileron travel should be applied into the wind as the takeoff roll begins. Gradually decrease the amount of applied aileron as the airspeed increases until only one-third to one-fourth aileron travel is applied at lift off.

During a crosswind takeoff it is advisable to raise the tail higher than normal to keep extra weight on the wheels and to keep the airplane on the ground until a higher airspeed is attained.

The tail should be raised until the aircraft is in level flight pitch attitude. When flying speed has been attained make the lift off in a smooth but very positive manner so that the aircraft will definitely remain airborne. The small amount of aileron into the wind will cause the aircraft to initiate a gentle turn into the crosswind. Allow this turn to continue momentarily until the proper crab angle has been established, then level the wings and stabilize the heading. Make any crab angle adjustment which may be needed so that the aircraft ground track is aligned over the departure runway. Crosswind takeoffs are usually associated with gusty surface winds thus aggressive and positive applications of flight controls are needed to assure a safe operation. A maximum crosswind component for takeoffs or landings should be set and not violated. As a general guideline I consider one-fourth of the stalling speed to be the upper limit. Good judgement must be applied and there are many variables to be considered.

Prominent variables are pilot experience and proficiency in T.W. operations and the characteristics of the airplane being operated.

WHERE TO LOOK DURING TAKEOFF

Looking directly over the nose will apply only to situations where the pilot is occupying the front seat and the aircraft is powered by a flat engine wherein the runway can be seen within about 800 to 1,000 feet ahead of the aircraft. In these situations looking over the nose may work from the beginning of the takeoff roll until lift off. Cessna 180 and 185's have a high cowling despite the flat engine installation.

The task of directional control is more difficult when the aircraft is equipped with a radial engine. The ability to see enough to maintain directional control will be limited whether the aircraft is open cockpit or cabin configuration. It is advantageous to utilize feel of the aircraft and peripheral visual cues to the greatest extent possible. The scheme which I prefer is to look out the left side of the windshield at an angle left of fuselage centerline. The line of vision will then project from the eyes past the left extremity of the engine and will provide an adequate view of the runway edge and the surrounding environment. As the takeoff roll progresses and the tail is raised the pilot's overall view will improve, in most situations, depending upon the aircraft such as a Cessna 195, Waco Cabin, Howard DGA-15P, Stearman, Waco UPF-7, etc.. The look point should be about 1,000 feet ahead of the aircraft. After a few takeoffs the proficient pilot will be able to manage directional and pitch control in spite of the restricted view of the runway.

THE NORMAL T.W. LANDING

The first step in preparing to land a T.W. aircraft should take place when the pilot first sits in the cockpit seat.

LOOK OUT PAST THE FRONT OF THE AIRCRAFT UTILIZING THE SAME VISUAL LOOK POINT AS FOR THE TAKEOFF PREVIOUSLY DISCUSSED. TAKE NOTE OF THE LOCATION WHERE THE HORIZON LINE MEETS THE COWLING OR ENGINE. THIS IS THE 3 POINT PITCH ATTITUDE WHICH SHOULD BE ATTAINED JUST BEFORE TOUCHDOWN.

This first look at the 3 point pitch attitude should involve 5 to 10 minutes of thoughtful orientation while looking full left and full right for general familiarization.

If the aircraft is open cockpit with no ring cowl, the 3 point attitude may be in relation to a cylinder or rocker box cover. If a ring cowling is installed a rivet, a row of rivets, a paint stripe, etc., may be the spot that meets the horizon line. Feel free to employ all the ingenuity necessary to establish a 3 point guide spot which will suffice. When the aircraft is a DH Beaver, Cessna 195, Staggerwing Beech or similar, another scheme is to note where the horizon line appears to cut through the cockpit. Since looking out left of the aircraft center line is preferred the 3 point guide spot may be even with the top to the attitude indicator as seen in peripheral vision. Additionally the 3 point guide point may be decal on the instrument panel a switch or whatever. Pick out something and refer to it on the landings. All these guide spots can be utilized during landings.

After 3 point landing proficiency is gained the necessity to refer to the 3 point guide point will continue. It then follows that night 3 point landings should not be attempted until skill is demonstrated in day landings.

The final approach, the flare and the look point for a 3 point T.W. landing are the same as for the tricycle gear aircraft except that the aircraft must be held off until the 3 point pitch attitude is attained. During the seconds just prior to touchdown the pilot visibility will be increasingly restricted as the nose of the aircraft is raised. During these few seconds and through touchdown and roll out the top priority is directional control.

During the first tailwheel landings directional control may seem to be overwhelming. However, through patience, good instruction and increased familiarity with the T.W. environment, the average pilot will soon master the beast!

Key clues are to be on the alert for any beginning directional deviation so that prompt and appropriate rudder inputs can be made. Stay off the brakes! Brakes may be needed to taxi and park but during the landing roll -- no brakes except in very unusual situations.

What to do if directional control and rudder authority deteriorate during the roll out? This situation is more likely to occur after the aircraft has slowed appreciably.

DIRECTIONAL CONTROL CAN BE SUPPLEMENTED OR REGAINED BY IMMEDIATELY APPLYING A BURST OF POWER. THE RESULTANT PROP BLAST WILL PROVIDE A SUBSTANTIAL INCREASE IN RUDDER AUTHORITY. APPLY THROTTLE UNTIL THE ENGINE SOUND APPROXIMATES THE SOUND OF ENGINE RUNUP.

These bursts of power should be short with the amount of power appropriate to the situation at hand. In desperate situations full power may be appropriate even including getting the aircraft airborne again.

IF A TAILDRAGGER IS ALLOWED TO DEVIATE MORE THAN APPROXIMATELY 10 DEGREES FROM THE RUNWAY HEADING LOOK OUT! THE LIKELY RESULT IS THAT THE INSURANCE COMPANY HAS BECOME A PARTNER AND THE AVIATOR IS MERELY ALONG FOR THE RIDE. AVOID THIS!

When the tail attempts to outrun the rest of the aircraft all hell breaks loose and the situation can turn real nasty. In tailwheel operation precise directional control is absolutely necessary. The pilot must remain alert as long as the wheels are turning! **The best never rest!**

WHEEL LANDINGS IN T.W. AIRCRAFT

To make a good wheel landing the approach should be the same as for a 3 point landing. The landing flare should be initiated closer to the runway. The flare must be planned so that the main gear wheels roll onto the runway at the bottom of the flare and while the tail is still up in approximately level flight attitude. The airspeed should decrease only slightly through the flare and until the touchdown. As the flare is begun a smooth and gradual power reduction should be initiated with the timing such that at touchdown the power is slightly above idle. After touchdown close the throttle and apply a barely noticeable amount of forward elevator to keep the wheels firmly on the runway. As airspeed decreases the tail will have a tendency to drop and the tailwheel will soon touch the runway. As the tailwheel touches the runway apply full aft stick or control wheel so that the tailwheel will be held firmly to the runway. **As the airplane slows down rudder authority will be decreased significantly.** The principal means of steering during the roll out is with positive applications of rudder, a tailwheel which is firmly on the runway and supplemental bursts of power, if needed. Directional control is most likely to be lost between the time that the tail is being lowered and until the airplane has slowed to taxi speed.

As wheel landing proficiency is increased the pilot can make power off wheel landings or variations to the power reduction as mentioned earlier. The power off wheel landing requires precise energy management so that the main gear wheels touchdown at approximate level flight attitude. If the airplane is held off or touchdown is delayed even an instant the tail will settle toward 3 point attitude. The sticky part is that on touchdown the aircraft will then skip back into the air.

A bad landing recovery and a 3 point landing will then be in order. A wheel landing which goes bad is best salvaged by making it into a 3 pointer or a go-around.

> THE *"HELL STRETCH"* IN EVERY WHEEL LANDING IS THE SHORT PERIOD DURING ROLL OUT AS THE TAIL IS BEING LOWERED AND BEFORE THE TAILWHEEL IS FIRMLY ON THE RUNWAY AND THE AIRPLANE HAS SLOWED TO TAXI SPEED. BE EVER VIGILANT.

CROSSWIND LANDINGS IN T.W. AIRCRAFT

Crosswind landings of either the 3 point or wheel landing type are virtually identical. Flying down the final approach should be done with a crab into the wind. As the flare is begun transition to the wing down into the wind/slip method. Maintain the fuselage aligned over and parallel to the runway center line by the use of rudder. Vary the amount of bank, i.e. wing down into the crosswind, so that the airplane remains over the runway. Too much aileron into the wind will cause the aircraft to slip off the runway upwind. Too little aileron will allow the airplane to be blown downwind from the runway.

As soon as the gear touches the runway during a crosswind wheel landing a fair amount of power should be added and adjusted as required throughout the roll out. During 3 point crosswind landings, a fair amount of power should be added as the tailwheel touches and adjusted as required throughout the roll out.

Ailerons, into the crosswind, must be **increased** as the aircraft slows prior to touchdown. Otherwise the aileron will begin losing effectivity shortly before touchdown which will allow the wing(s) to roll toward level flight. When this happens the airplane will be blown off the runway in the downwind direction.

Hold the stick or control full aft with aileron into the crosswind through all roll outs. Taxi into the ramp with flight controls positioned appropriately for the wind directions which prevail. Use brakes only when needed.

Wheel landings are easier to accomplish in a heavier aircraft such as a DC-3 or Twin Beech. Light airplanes tend to skip back airborne unless the wheel landing is very smooth. The preference for wheel or 3 point landings vary from pilot to pilot and depend on the job being done and the aircraft being operated. My preference is do 3 point landings in single engine T.W. aircraft during most conditions. In conditions of bad crosswind and/or heavy turbulence I usually opt for a wheel landing where **I can nail it on** under full control. Once the airplane is under control directionally the heavy work begins, namely maintaining directional control as the aircraft slows and the tail is lowered and until some help is derived from tailwheel steering.

COULD THESE ACCIDENTS HAVE BEEN PREVENTED?

A student pilot was flying a Piper Super Cub during a local solo flight. He buzzed town at a very low altitude then pulled up sharply into a *"hammerhead"* maneuver. The aircraft then descended nose down into the ground. The student pilot was killed and the aircraft was destroyed.

In northern Wyoming a high time pilot was landing a J-3 Cub on a private strip which was covered with 5 or 6 inches of dry powdered snow. The aircraft touched down then pitched forward and nosed over. The pilot discovered that the wheels were frozen and would not rotate. Before the flight, the pilot had taxied through 6 to 8 inches of snow and had used his brakes. The pilot was uninjured.

In Alaska a pilot was making an approach to land on a highway near the lodge which he operated. This operation was a common practice. At low altitude the aircraft appeared to stall and crashed onto the highway. The airplane burst into flames and the three occupants were killed.

In California a Cessna 180 was landing during a steady crosswind component of 9 knots. A gust was encountered at touchdown and the aircraft ground looped. One wing received substantial damage. The occupants were not hurt. The pilot had more than 2,000 hours in Cessna 180 aircraft.

In Alaska a Cessna 185 Skywagon ground looped when directional control was lost during the takeoff roll. The flight instructor, student pilot/owner and a passenger were not injured. The aircraft suffered substantial damage.

Taildragger pilots appear to be involved in a disproportionate number of incidents involving attempted aerobatics without having had instruction. Additionally, the taildragger pilots get involved in flying visually into IFR conditions, night flying without proper training and the long list of self destructive activities which are attributed over and over to the general pilot population.

THE BOTTOM LINE IN TAILWHEEL AIRCRAFT FLYING IS TO MAINTAIN PRECISE DIRECTIONAL CONTROL DURING TAKEOFF AND LANDING OPERATIONS.

CHAPTER BRIEFING NOTES

This chapter is intended to be an overview of high country mountain flying in the western U.S. The major focus is on the performance limitations which apply during high altitude and high density altitude operations. A common thread in most mountain flying accidents is that the pilot is expecting the aircraft and engine to perform beyond their capabilities.

This chapter assumes that operations are to and from regularly used high country airports with paved runways. There is no attempt to address the special techniques and skills which are unique to **bush** type operations which use unimproved and remote strips. All operations are to be conducted in day VFR conditions only. No IFR and no night operations -- both are very challenging and potentially dangerous.

In North America there are 82 peaks with elevations of 14,000 feet or higher. 54 of these peaks are in Colorado. Thus the focus is on the Denver area, where I have lived since 1955, and the Colorado High Country west and southwest of Denver. A Denver sectional chart will be beneficial here because there is reference to various airports and passes which may be of interest to the reader.

Some of the material in this chapter was developed by me in 1990-91 and donated to Colorado Pilots Association to be the basic content of their copyrighted text entitled **MOUNTAIN FLYING**. Subsequently the Colorado Pilots Association granted me specific approval to use materials from this text.

The text **MOUNTAIN FLYING** is the reference manual for the widely acclaimed Colorado Pilots Association mountain flying course.

The ground school has been taught in many states outside Colorado. The optional mountain cross country flight training is conducted in the trainee pilots aircraft. The flight training is usually conducted from Centennial Airport which is located at the south edge of the Denver metro area. The training route begins at Centennial Airport, across Corona Pass direct to Kremmling and land. Then direct Eagle and direct Glenwood Springs Airport and land. Then direct Aspen and land. Next to Leadville via Reudi Reservoir and Hagerman Pass or via Eagle, Minturn and Tennessee Pass. Leadville is the highest public airport in North America at an elevation of 9,927 feet. Fly the downwind leg at 10,927 feet! From Leadville the route is via Buena Vista and 25 NM north of Colorado Springs to Centennial Airport.

The first CPA mountain flying ground school was taught in it's entirety by fellow CFI Bruce Hulley beginning in 1986. The new mountain flying course came on line in June 1991. It was developed principally by fellow CFI's Robert Henley, Bill Standerfer, Bill Marshall and myself.

THIS CHAPTER ENTITLED MOUNTAIN FLYING - WHAT YOU NEED TO KNOW -- IS NOT INTENDED TO QUALIFY A PILOT TO CONDUCT MOUNTAIN FLYING OPERATIONS. NO PILOT SHOULD ATTEMPT MOUNTAIN FLYING OPERATIONS UNTIL AFTER HAVING A THOROUGH MOUNTAIN FLYING GROUND SCHOOL AND A MOUNTAIN XC CHECK OUT BY A FLIGHT INSTRUCTOR WHO HAS EXTENSIVE MOUNTAIN FLYING EXPERIENCE.

This chapter **is** intended to develop the readers understanding of major factors which impact the performance and operation of general aviation aircraft when engaged in mountain flying operations in the high country of the western U.S.

Note that FIGURES 15-2 and 15-3 are intended to illustrate typical performance numbers. These figures do not relate to any specific make and model of aircraft. Check the performance charts in the POH for each aircraft you fly.

MINIMUM REQUIREMENTS FOR SAFE MOUNTAIN FLYING

PILOT - Private License, 200 hours total time, current FAA medical certificate, current Biennial Flight Review and 10 landings in the past 30 days in the aircraft being operated.

AIRCRAFT - Equipped with an engine rated at 180 H.P. or more. Light twins need engines each rated at 225 H.P. or more. On board equipment to include a current and certified ELT, mode C transponder and a VHF Nav-Com.

WEATHER - Minimum visibility is 15 miles. Visibility is routinely 20 to 50 miles in the high country. A ceiling high enough so that there is at least 2,000 feet of clearance above the highest pass or terrain enroute. This clearance is to provide for the ceiling to be 1,000 feet above the aircraft and for the aircraft to be 1,000 feet above the highest terrain.

Most high country weather reporting stations are located in valleys or in flat land areas. For example in order to safely cross Corona Pass - 11,680 feet, and located about 40 miles west of Denver, the ceiling at prominent Denver airports must be approximately 8,000 feet and the ceiling at Kremmling which is west of the pass must be about 6,400 feet. See FIGURE 15-1.

The winds aloft at 9.000 and 12,000 feet must not exceed 30 knots. Otherwise there is high risk of encountering very turbulent air and updrafts and downdrafts of 800 to 1,000 FPM or more.

These downdrafts will far exceed the climb capability of the aircraft being operated. Excessively strong winds aloft, of 50 to 75 kts., occur during the months of January, February and March. Winds aloft usually exceed 100 kts. on several occasions during each winter.

This is not to say that general aviation aircraft can not be operated in the high country during winter months. However, days with suitable winds aloft and suitable area weather will be scarce.

GROSS WEIGHT - The aircraft gross takeoff weight should not exceed 90 percent of the FAA certificated maximum gross weight. This 10 percent decrease in gross weight will allow the rate of climb and the service ceiling to be increased significantly during high density altitude operations.

SURVIVAL GEAR - Adequate survival gear depends upon the time of year and the route being flown. Obvious needs are warm clothing, water, food, medical kit, matches, flashlight, and a small tarp or plastic cover, a sturdy knife and several feet of strong cord. A hand held VHF NAV-COMM would be invaluable. This list is not intended to be a complete list but rather a listing of items usually included in a survival kit. The pilot in command should assemble a survival kit which contains all items deemed appropriate for the mountain cross country flight being planned.

FUEL ON BOARD - The fuel quantity should provide for all the enroute burn out plus a minimum reserve of one hour of fuel upon arrival at the destination. My preference is to have a fuel reserve of one hour and 30 minutes.

It is easy to encounter enroute headwinds which are stronger than expected. You may need to shuttle climb to gain sufficient altitude for a safe pass crossing.

An extra 30 minutes of fuel could be consumed in the two aforementioned situations. I suggest adding two minutes of time and fuel consumption for each 1,000 feet of climb between the departure airport and highest planned enroute altitude.

SECTIONAL CHARTS AND FLIGHT LOG - Have on board sectional charts for the route and the area out to 100 miles each side of the planned route. A detour maybe required or it may be necessary to divert to an alternate airport. A flight log is needed to document the progress of the flight and the fuel consumption and quantity remaining.

DENSITY ALTITUDE

Density altitude is pressure altitude adjusted for temperature. I like the practical definition which states: **DENSITY ALTITUDE IS THE ALTITUDE AT WHICH THE AIRPLANE THINKS IT IS OPERATING REGARDLESS OF THE ALTIMETER READING.**

> THE EFFECTS OF DENSITY ALTITUDE ARE MOST SEVERE WHEN AN AIRCRAFT IS BEING OPERATED ABOVE ABOUT 4,000 FEET MSL, AT OR NEAR MAXIMUM GROSS WEIGHT AND WITH THE O.A.T. HIGHER THAN STANDARD.

During the summer months the O.A.T. is frequently standard plus 30 degrees F. This temperature will equate to having a density altitude of nearly 3,000 feet higher than the actual altitude. When the O.A.T. reaches standard plus 40 degrees F operations from airports above 4,000 feet elevation will suffer severe performance penalties. Takeoff distances will be much longer and the initial rate of climb may be alarmingly low.

The following table shows density altitudes versus airport elevations when the O.A.T. is at standard plus 40 degrees F in each example.

AIRPORT ELEV		STD.TEMP	STD.+40°F	DENSITY ALT.
Centennial	5,883 FT.	38°F	78°F	8,500 FT.
Aspen	7,815 FT.	32°F	72°F	10,300 FT.
Leadville	9,927 FT.	23°F	63°F	12,400 FT.

It is interesting to note that the takeoff ground roll at Leadville will be approximately 2.75 times as long as at sea level - if the Leadville temperature is 63 degrees F. You may see a POH which does not state climb performance at the Leadville density altitude of 12,400 feet. Remember the airplane thinks it is at 12,400 feet as the takeoff is begun from the 9,927 foot elevation of the runway. What to do? Consult the long used KOCH chart and make your most reasonable calculation of performance.

The following is a list of the effects, on aircraft performance, caused by high altitudes and/or high density altitudes. Performance is much degraded at high elevations regardless of the temperature. Check the POH.

- Takeoff distances are dramatically increased.
- Landing distances are increased but at a lesser amount than the takeoff distances.
- Rates of climb are reduced a great deal.
- Service ceilings are at a much reduced **actual altitude.** For example a Cessna 150 has a service ceiling of 12,650 feet at standard temperature. However, at an O.A.T. of standard plus 40 degrees F thc maximum actual altitude of the Cessna 150 would be approximately 9,400 feet.

The difference of 3,250 feet between the service ceiling and the highest attainable actual altitude represents the density altitude penalty which is commonplace on a hot day.

- A normally aspirated engine will lose approximately 3 percent of rated power for each 1,000 feet above sea level. The following power output comparisons for various altitudes are accurate enough for operational purposes.

- Power output comparisons with a 285 HP engine:
At 5,900 feet (DEN area) 23.5 inches MP - 236 HP.
At 10,000 feet (Leadville) 19.5 inches MP - 200 HP.

- Power output comparisons with a 180 HP engine:
At 5,900 feet (DEN area) 23.5 inches MP - 149 HP.
At 10,000 feet (Leadville) 19.5 inches MP - 126 HP.

The above examples are to show the amount of power loss at higher altitudes due to the decrease in full throttle manifold pressure. Humidity and temperature are factors in the power loss also -- but not addressed here because the operational impact is minimal. Note that full throttle takeoff power at Leadville will be only about 62 to 65 percent of sea level takeoff power. This is actually a conservative **cruise power** setting.

Now let us review the impact of the increase in true airspeed with altitude above sea level. The result is much increased takeoff distances at high altitude airports. This increase in TAS amounts to approximately 2 percent for each 1,000 feet above sea level. Let's look at a Leadville takeoff assuming the lift off speed is 75 kts. IAS. We will assume a zero surface wind for this example.

The true airspeed correction is 2 percent times 10 which represents the Leadville elevation in thousands of feet. We have 20 percent of 75 or 15 kts. TAS as a correction to add to the 75 kts. IAS at liftoff. Thus at liftoff the aircraft will be traveling down the runway at 90 kts. ground speed at the moment the airspeed indicator reads 75 kts. In this example we are accelerating an **extra 15 kts. ground speed and doing it with cruise power!** Therein lies the answer to why takeoffs -- when high and hot -- require lots of runway length. See FIGURE 15-2.

- Turbo-charged engines - can maintain sea level manifold pressure to around 10,000 feet or higher. Thus the takeoff roll will be shortened to slightly less than if the same aircraft were powered by a normally aspirated engine. The main advantage of the turbo-charged engine is that the rate of climb will be near the sea level rate to a higher altitude.

- Twin engine takeoff performance - It may come as a surprise to learn that light twins usually require a longer takeoff roll than their single engine cousins. The reason is that the light twin liftoff speeds are 10 to 15 kts. IAS higher than a comparable single engine aircraft. At the Leadville airport -- A Cessna 182, Mooney or Bonanza will usually lift off in around 2,500 feet of ground roll. In the same conditions namely warm day and zero wind the light twins, such as Cessna 310's and Beech Barons, will require a ground roll of 3,200 to 3,400 feet. See FIGURE 15-3 for light twin accelerate/stop distances and climb performance with engine failure shortly after lift off.

The above examples are rough generalizations based upon my experience as a mountain flying CFI and pilot. The best rate of climb IAS decreases with altitude and must be adhered to if maximum performance is to be derived.

Check the POH for a chart on climb speeds versus altitude. If there is no chart in the POH an approximation is that the IAS for climb at 10,000 feet MSL will be 5 to 7 kts. IAS less than the IAS for climb at sea level. Failure to adhere to the proper climb IAS will result in the service ceiling being reduced by 1,000 to 1,500 feet.

ROUTE SELECTION

The route to be flown should be over lower terrain even if some detouring is involved. Long airport to airport direct flights will often require a lot of operations at or above 15,000 feet. Take time to now review FAR 91.211 which specifies the requirements for usage of oxygen by pilot and passengers. Check the Denver sectional chart and note that the dark brown contour coloring begins at the 9,000 foot elevation - thus the darker the color the higher the terrain. Select a route which follows valleys and lower terrain between check points. Pass crossings, in the high country, frequently require crossing altitudes of 11,000 to 13,500 feet. Cruising altitudes, between passes and very high terrain, will range from 8,500 to 11,500 feet.

NAVIGATION

The recommended basic method of mountain XC navigation is dead reckoning. VORs, DME and radar coverage are generally unavailable except at very high altitudes. The flight plan should be made with point to point magnetic courses laid out and with distances and estimated times between check points. During the flight the course should be adhered to closely while checking the track by pilotage with the sectional chart. Attempting to navigate by pilotage alone involves a high risk of confusion and disorientation. A Denver sectional chart may be helpful as this chapter progresses.

The safest and most precise method is to note that the course from A to B is for example, 258 degrees and the distance is 58 NM. Depart checkpoint A flying a heading of 258 degrees. If a drift correction is needed apply it. You can pick out distant on course visual references by aiming the airplane momentarily on 258 degrees and note the visual reference to be flown to. Then apply any drift correction which is appropriate. The visual reference may be a mountain peak, lake, bare spot, etc.

THE AIRPLANE IS BEING FLOWN TOWARD *"LOST"* IF, WHILE ENROUTE, THE HEADING IS ALLOWED TO DEVIATE MORE THAN A FEW DEGREES FROM THE BASIC HEADING. HANG ON TO THE DESIRED HEADING.

ENGINE START AND TAXI

High altitude engine starts should be done using the normal engine start procedure. As soon as the engine begins to run check oil pressure, set RPM at 1,000 or less and immediately lean the mixture by moving the mixture toward lean until RPM increases slightly and the engine runs smoothly. The mixture control will usually be moved aft about one-half inch. The reason for the leaning it to prevent fast and excessive build-up of lead on the spark plugs during taxi operations.

RUN UP

Cont. IO-550 Mix. RICH for Run Up & All T/O & Ldgs.

Upon reaching the run up area enrich the mixture slightly and carry out a normal run up whether the engine is turbo-charged or normally operated.

For a normally aspirated engine follow the run up by enriching the mixture slightly and applying full throttle. Next move expeditiously to adjust the mixture for maximum power by EGT gauge, fuel flow meter or check the POH as appropriate for the engine being operated.

Next smoothly reduce the RPM to 1,000 or less. The mixture will now be set for takeoff power. Check your POH. Get training in your plane before attempting this one on your own.

For the turbo-charged engine conclude the run up and reduce the RPM to 1,000 or less while awaiting takeoff. When ready to begin the takeoff roll move the mixture control to full rich as takeoff power is applied. The reason here is that the turbo-charged engine will develop rated manifold pressure even at high altitude airports thus full rich mixture is required.

TAKEOFF AND INITIAL CLIMB

Flaps at zero is recommended for high country takeoffs. The reason is that the extra drag of flaps coupled with decreased engine power will seriously degrade climb performance. Move into takeoff position as near the departure end of the runway as is practical. Hold the brakes, apply takeoff power, make a quick power check - RPM, MP and fuel flow as appropriate - then release the brakes. During the takeoff roll make another power check prior to beginning the rotation for lift off. The rotation should be started just prior to V_{xse} for twin engine and V_x for single engine aircraft. Rotate to the predetermined nose up pitch attitude for initial climb. Keep hand on the throttle(s). Climb at V_y until reaching 1,000 feet above the airport or higher if desirable. Remove hand from the throttle(s) only to operate the gear switch or other necessary controls. See FIGURES 15-4 and 15-5.

ENROUTE CLIMB

Transition from initial climb to enroute climb by lowering the nose slightly and when the desired enroute climb IAS is attained set climb power.

Check fuel flow or mixture settings each 2,000 feet of climb beginning at the elevation of the departure airport. For maximum climb performance adhere to the V_y IAS which is appropriate for the altitudes transited during climb. Check the POH for speeds.

It is extremely important to monitor the nose up pitch attitude via the pitch attitude indicator throughout the climb. High country visual cues are very confusing since they consist of valleys, mountains, plateaus and a variety of terrain features not normally seen during flat land operations.

Monitor the rate of climb instrument throughout the climb. The importance is that a down draft may be encountered in glassy smooth air -- even in the down flow of a mountain wave!

When one of these downdrafts does occur it usually goes undetected for several seconds if the pilot is relying solely on seat of the pants feel. A continuous monitoring of the climb rate is very important throughout the climb.

Monitor and control the heading throughout the climb. This is especially important when climbing toward a pass which can not be recognized until at or near the pass elevation.

PASS CROSSING

The salient information needed in advance is the elevation of the pass. Add 1,000 feet to the pass elevation to derive the minimum pass crossing altitude. This altitude must be attained at least 3 miles before reaching the pass. If the pass crossing altitude is not reached at or before the 3 mile point then a shuttle climb is required. See FIGURE 15-6.

The shuttle climb is accomplished by continuing the climb while turning so as to parallel the mountain ridge. Fly this heading for one or two minutes then turn away from the ridge and continue making a race track pattern until the pass crossing altitude is reached. Shuttle climb is frequently required when crossing Corona Pass which is about 35 miles west of Denver. If there is insufficient room to shuttle climb then an option may be to try circling with a shallow bank while climbing. Another option is to climb while retracing, in reverse, the ground track which was toward the pass. The last option may be the best of all and that is to fly to another area where a crossing can be accomplished under less challenging conditions. When all else fails avail yourself of my favorite alternate airport -- the departure point! Pass crossing should be at a 45 degree angle if surrounding terrain permits.

Some passes are at the top of long V-shaped valleys. This type of pass can be identified by studying the contour lines on the sectional. Look on the Denver sectional chart and locate the town of Minturn which is about 25 miles east of the Eagle Colorado airport and at a sharp bend in the 4-lane I-70. Looking SSE on a heading of 150 degrees we see the V-shaped valley leading up to Tennessee Pass elevation 10,424 feet. To safely cross this pass flying southbound we should climb to the pass crossing altitude of 11,500 feet before starting up the valley. If we enter this valley at low altitude we may find that the highway is climbing faster than the airplane is capable of climbing and there is not enough room to turn around should the need arise. This is the scenario which traps and average of 6 or 7 pilots each year here in Colorado. These situations frequently lead to loss of life. Most are serious crashes. Infrequently the pilot is able to land on the highway. The Highway Patrol has yet to see any humor in such landings.

Tennessee Pass does not have the room to accommodate a 45 degree crossing track as does Corona Pass, thus I usually go straight over this one. When in a V-shaped pass, with lots of altitude, the flight can take advantage of the updrafts that usually occur on the windward/updraft side of the valley. A turn to reverse course can also be completed safely should the need arise.

A third general type of high country pass is exemplified by Wolf Creek Pass about 20 miles NNE of Pagosa Springs in southern Colorado. This pass road turns abruptly toward the west at the pass. If a pilot was to follow the pass precisely it would necessitate very sharp turns in the pass crossing area. My preference is to go directly over the highest terrain in the pass area. For this reason I like to add 2,000 feet to the 10,857 elevation at Wolf Creek Pass. If the air is turbulent I add an extra 500 feet!

If a down draft is encountered, while on the track for crossing any pass, an immediate turn must be started to move toward lower terrain. This escape maneuver is easier and quicker if the crossing track is at a 45 degree angle. For pass crossings which are to be made essentially straight-over the crossing should be along one side of the valley so that the escape maneuver will be to turn toward the lower part of the valley.

During the winter of 1993 my friend Bill Marshall and I were enroute westbound from Denver to cross Corona Pass. We were in Bill's airplane. Bill was taking pictures while I flew the airplane. The winds at 9,000 and 12,000 feet were reported at 230 degrees 30 kts. It was necessary to utilize a shuttle climb to get to the pass crossing altitude because we encountered several very low rates of climb on the way up from the front range area.

We began the pass crossing route on a heading of about 315 degrees. The pass elevation is 11,680 feet. I had added the minimum 1,000 feet of clearance to the pass elevation plus a little additional altitude for insurance. We began the pass crossing route at 13,100 feet and were moving slowly, perhaps 60 kts. ground speed. The wind appeared to be from a more westerly direction than forecast.

When we were about 2 miles/2 minutes from the top of the pass the rate of climb indicator abruptly indicated **600 FPM DOWN!** In a heartbeat I began a turn to the east and away from the pass. This rapid descent continued until we were well below the pass elevation. Had we not been approaching the pass at a 45 degree angle with extra altitude and had I not been watching the rate of climb/descent we almost surely would have slammed into the 40 foot snow bank on the east side of the pass. This was the first time I had the need to use the escape maneuver. I was prepared and it paid off handsomely. We scrubbed that flight and returned to base.

ROUGH AIR SPEEDS

It is essential that rough air speeds be fully understood and that the numbers for the aircraft being flown are at hand. Check the POH for the maneuvering speed. First off this IAS is for the aircraft at certificated gross weight. If the aircraft is flying at a lesser weight the maneuvering speed will be less. To be safe the pilot then needs a **heavy weight** maneuvering speed and a **light weight** maneuvering speed.

Once these two maneuvering speeds are known it is then reasonable to subtract a few knots IAS because in moderate to severe turbulence the maneuvering speed should not be exceeded.

As we all know the airspeed can fluctuate wildly in turbulence. I suggest using a rough air speed which is 10 kts or so below the maneuvering speed. See FIGURE 15-7.

ENROUTE CRUISE

This phase of flight consists of navigation, making position reports and the monitoring and managing of the flight. In some areas you may unable to contact the FSS direct or via a RCO or VOR facility. In such a situation try to contact **FLIGHT WATCH ON 122.0** because in the high country the FSS antenna and FLIGHT WATCH antenna are not always co-located.

You may be out of range of an FSS facility but virtually on top of a FLIGHT WATCH remote. The FLIGHT WATCH specialist will be glad to assist you if requested to do so. I have used FLIGHT WATCH, for position reports, on many occasions.

THE DESCENT FROM CRUISE

Unusual descents are frequently required for airports which are located in deep valleys. In many of these situations the traditional traffic pattern cannot be flown due to restricted clearance surrounding the runway. Glenwood Springs, CO. airport is a good example, especially if the route from the east along the canyon is flown at 12,500 feet MSL. Make frequent reports of altitude and position during all descents, approaches, and departures from non-tower airports.

In this example slow the aircraft in level flight, extend the gear, if a retract, extend about half flaps, set RPM at about 1500 or MP at about 15 inches. Allow the aircraft to descend at around 500-600 FPM.

The IAS will be slightly above approach speed so that any turbulence will be managed with a minimum of passenger discomfort and airframe strain. Enrich the mixture slightly each 2,000 feet during the descent. While in the descent look over the area surrounding the runway and decide on the safest go-around track in case a go-around becomes necessary. In the case of the Glenwood Springs airport, landings are on runway 32 and takeoffs are opposite direction on runway 14. The go-around track is to proceed northwest following the river and over downtown then turn left at the 4-lane highway. Continue climbing westbound until there is enough room to turn around or divert west to the Rifle, CO. airport which has clear approaches.

APPROACH AND LANDING

When about 1,500 feet above the airport apply full power momentarily and adjust the mixture. This will set the mixture for a go-around if needed with a normally aspirated engine. If the engine has a carburetor heat control then apply carburetor heat for a few seconds at about 1,000 ft. AGL on the approach and then return the control to COLD. Do not fly the approach with the carburetor heat on HOT as this will cause a loss of power and an excessively rich mixture.

If the engine is turbo-charged I suggest momentarily applying climb power at around 1,500 feet AGL and adjusting the fuel flow. The mixture will then be suitable for the approach. In case of a go-around go to full rich as takeoff power is applied.

If the mixture is set at full rich at higher than about 4,000 feet MSL, during an approach, the excessively rich mixture will cause the engine to run rough. In some cases the engine will actually die.

This applies to both normally aspirated and turbo-charged engines. Beware that POH check lists call for full rich for landing -- do not do this on high altitude approaches!
For guidance on the approach refer to FIGURES 15-8 and 15-9 which are repeats of the PRECISION VISUAL APPROACH. For the Glenwood Springs approach begin a long straight in approach from the south to runway 32. The aim point should be about 300-400 feet beyond the threshold. This will provide a touchdown about 800-1,000 feet from the threshold. There will then remain sufficient runway for a normal roll out. Runway 32/14 is 3,300 feet long. Avoid the lines and trees on the edge of the river. Leave the mixture leaned while taxiing to the ramp.

GO-AROUNDS

Reasons to initiate a go-around include: excessive turbulence during the approach, excessive cross wind component, a disabled aircraft on the runway, and the approach too high or too fast. The bottom line is that the approach must be on profile, on speed and stabilized at 100 feet AGL. If not, it is go-around time. Do not attempt to land out of a poor approach because the result is almost always a bad landing -- either a short or long touchdown, or a departure beyond the far end of the runway.

EMERGENCY LANDINGS IN THE HIGH COUNTRY

If an engine fails immediately begin a turn toward lower terrain while maintaining altitude until the airspeed decreases to best glide speed. See the POH. Attempt an inflight engine re-start per the POH procedure.

If the engine restart is not successful declare an emergency on tower or UNICOM if in the vicinity of an airport, otherwise 121.5, FSS, or ATC center frequency where you may be heard by an airliner overhead. Proceed toward the most promising landing area. The first priority is to save the occupants from injury. Expect the airplane to be severely damaged. A landing at a lower elevation site offers the advantage of being nearer to habitation, vehicle traffic, water, warmer temperatures and less wind.

In preparing for the landing assure that all occupants have seat belts and shoulder harness secure and adjusted properly. Brief passengers on the operation of the emergency exit if one is installed. Direct the passengers to move away from the aircraft after touchdown in case a fire should break out. If there is an ELT switch in the cockpit turn it on during the approach. Unlatch the cabin door(s) as appropriate. The doors will trail open about 3 inches at the aft edge. Shortly before touchdown boost pump(s)-OFF, fuel selector(s)-OFF, ignition switch(s)-OFF, electrical master switch-OFF. Touchdown at the slowest safe airspeed. My preference for airplane configuration is gear down and full flaps. This is a decision each pilot must make considering the situation at hand.

AFTER THE EMERGENCY LANDING

Check the status of the occupants and carry out any needed first aid. It is very important that all persons keep dry and warm.

ASSURE THAT THE ELT IS OPERATING. IT MAY BE NECESSARY TO TURN THE SWITCH TO *"ON"* AT THE ELT UNIT. DO NOT RELY ON THE ELT IMPACT SWITCH TO OPERATE IN EVERY INSTANCE.

Civil Air Patrol statistics indicate that the average time to rescue is 13.7 hours if the ELT is transmitting and 55.6 hours if the ELT is not transmitting.

Arrange the best possible shelter utilizing the aircraft if advisable. Leaving shelter to walk out for help is generally very dangerous. This should be attempted only with good visibility during daylight and by a very fit person who is warmly clothed. Additionally, if a known point of contact is not nearby, the safest and best bet is to remain with the shelter and await rescue.

The rescue can be expedited by making an early radio contact with another aircraft. Here is where it pays to have on hand the **JEPP** or **NOS IFR LOW ALTITUDE ENROUTE CHART** for the area. Try transmitting in the *"blind"* on one or more of the nearby enroute ATC frequencies listed on the IFR chart. These frequencies are being monitored by airline, corporate, and private pilots. Any of these pilots will respond to your call and willingly assist by making contact with appropriate authorities. Here you can see the importance of knowing where you are at all times. A hand-held **VHF NAV-COMM** is invaluable in a situation such as this. It is advisable to make a transmission each 5 minutes or so until a contact is made.

COULD THESE ACCIDENTS HAVE BEEN PREVENTED?

A Learjet was **on a VFR** night approach into Eagle, CO. when it collided with high terrain which surrounds the airport. The airplane was destroyed and all occupants perished.

A Piper Cherokee crashed near the top of Monarch Pass, west of Salida, CO. The PIC was a CFI from out of state.

Two off duty airline pilots crashed during low altitude sight seeing in a four place retractable aircraft. The location was in the vicinity of a ski area. The accident occurred in day VFR conditions during the winter. One pilot was found dead in the aircraft. The other was found frozen to death after he tried to hike out for help. Both men were dressed in street clothes.

A Bonanza was departing Centennial airport when it failed to get airborne and ran off the departure end of the runway. The airplane was thought to be overloaded and operated with the mixture full rich. The pilot was from a flatland location.

Ceiling Needed to Cross Corona Pass

- Route of flight is Denver-Centennial Airport (APA) to 6 miles west of JEFFCO Airport, Corona Pass to Kremmling (RLG).

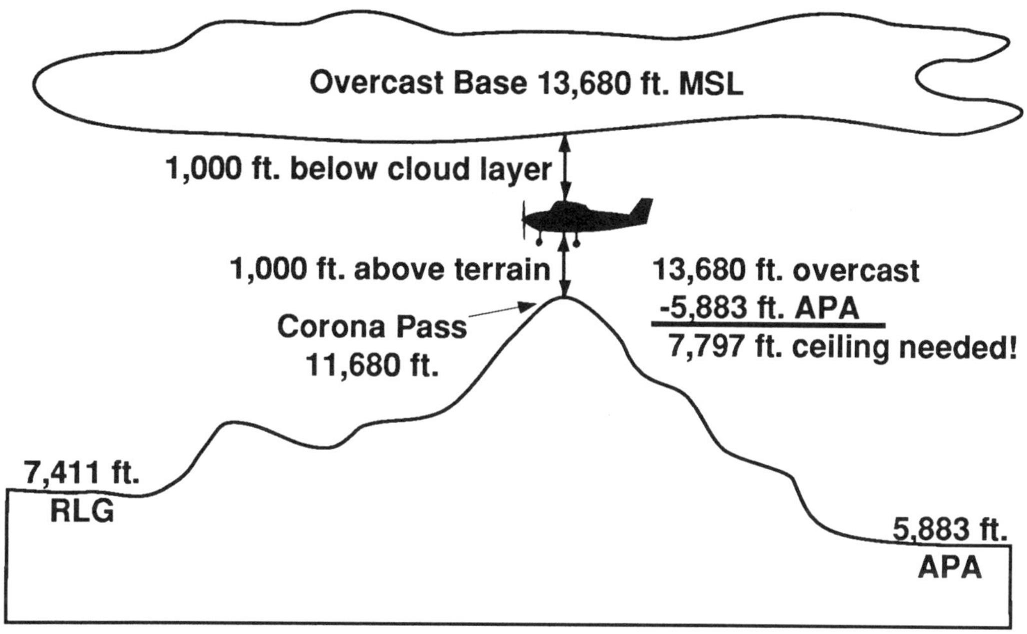

- Most weather reporting stations and airports are in valleys.
- Ceilings of 5,000 to 8,000 ft. are necessary for safe high country mountain flying

Figure 15-1

Typical S.E. Aircraft Performance

These charts focus on the differences in performance between an aircraft at maximum FAA certificated gross weight and at 90 percent of the maximum gross weight. The O.A.T. is standard pluse 40 degrees F.

Ground run to lift off

Airport Elevation	Sea Level	6,000 ft.	8,000 ft.	10,000 ft.
Max Gross Wt.	1,260 ft.	1,880 ft.	2,810 ft.	3,275 ft.
90% Max. Gr. Wt.	1,000 ft.	1,610 ft.	2,120 ft.	2,650 ft.

Rate of Climb vs. Gross Wt.

Altitude	Sea Level	9,000 ft.	12,000 ft.
Max Gross Wt.	700 FPM	400 FPM	270 FPM
90% Max. Gr. Wt.	875 FPM	500 FPM	400 FPM

Service Ceiling vs. Gross Wt.

AIRCRAFT WT.	SERVICE CEILING
Max. Gross Wt.	13,200 ft.
90% Max. Gr. Wt.	15,300 ft

Figure 15-2

Typical Light Twin Performance
Max. Gross Wt. vs 90% Max. Gross Wt.

Accelerate/stop distances vs. gross weight

Field Elevation	5,100 lbs.	4,500 lbs.
6,000 ft. @ 80 °F	5,000 ft.	4,600 ft.
8,000 ft. @ 70 °F	5,600 ft.	5,100 ft.
10,000 ft @ 63 °F	6,100 ft.	5,600 ft.

Typical light twin climb or descent following an engine failure shortly after lift off

Field Elevation	5,100 lbs.	4,500 lbs.
6,000 ft. @ 80 °F	May hold altitude	Climb 150 fpm
8,000 ft. @ 70 °F	Descend 100 fpm	Climb 125 fpm
10,000 ft @ 63 °F	Descend 200 fpm	May hold altitude

O.A.T. is approx. STD plus 40 °F

Figure 15-3

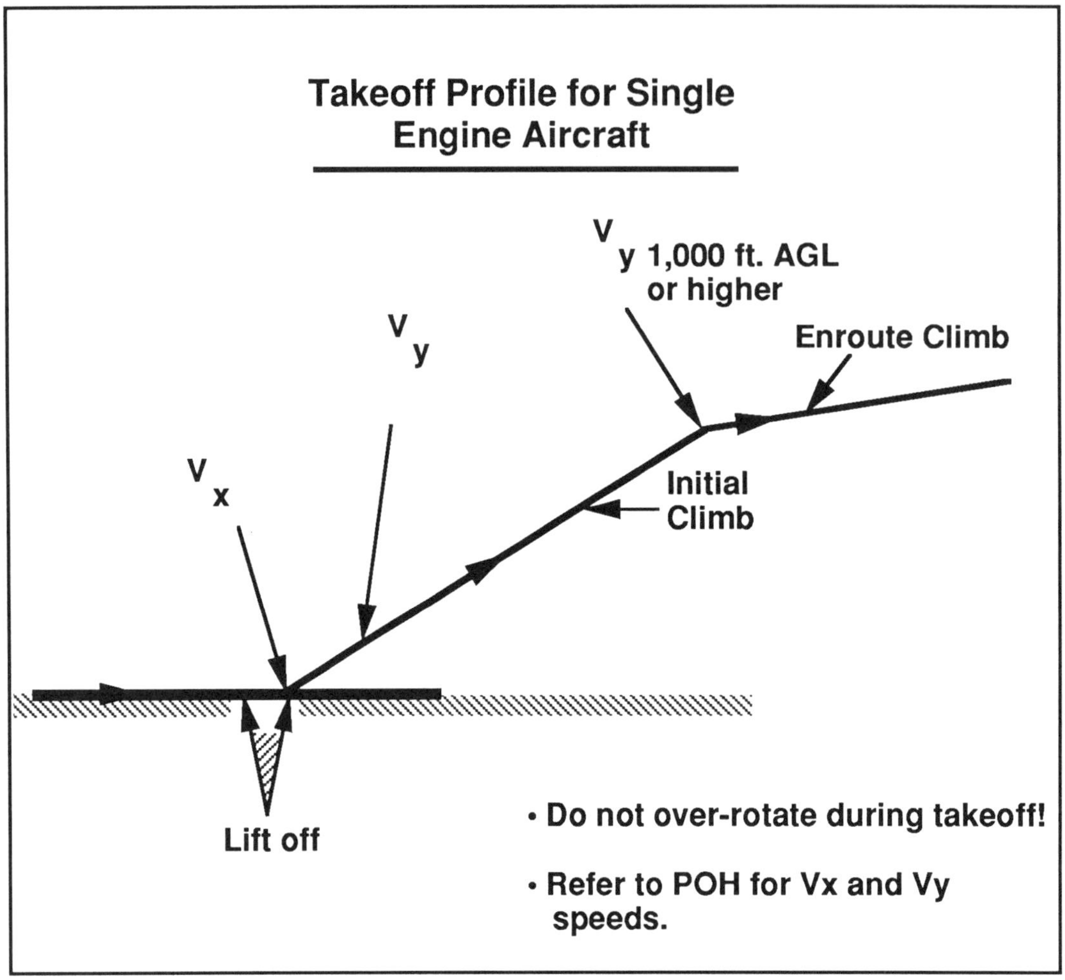

- Begin rotation at the appropriate IAS.
- Rotate to the pitch attitude for climb at Vy (Note this pitch attitude for the airplane you fly).
- Vy will usually be attained at about 75 ft. AGL.
- Note the normal rate of climb during each initial climb. If the climb rate is significantly below normal, check the power setting, gear, flaps, etc.

Figure 15-4

Takeoff Profile for Twin Engine Aircraft

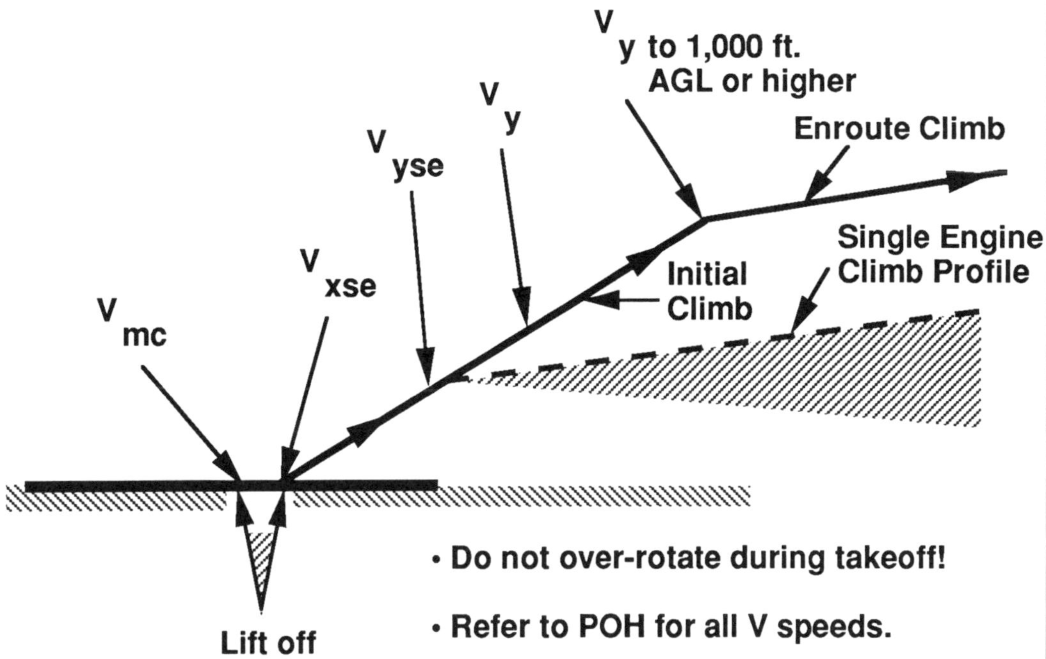

- Do not over-rotate during takeoff!
- Refer to POH for all V speeds.
- — — — Portrays S.E. climb out at Vyse (Blue Line IAS) after a power failure - if S.E. climb capability exists.

- Begin rotation so that lift off will occur at Vxse.
- Rotate to the pitch attitude for climb at Vy (Note this pitch attitude for the twin you fly).
- Vyse will usually be attained at about 50 ft. AGL.
- Note the normal rate of climb during each initial climb. If the climb rate is significantly below normal, check power settings, gear, flaps, etc.

Figure 15-5

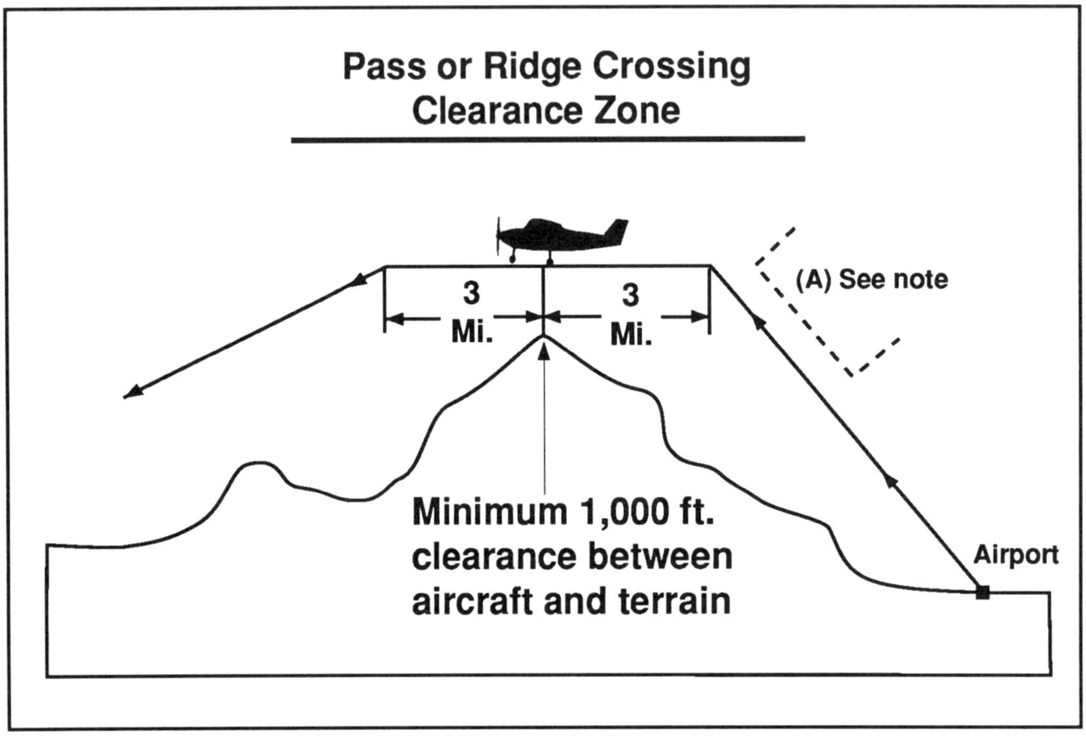

(A) Execute a shuttle climb, at any time, if you are unable to comply with the pass crossing altitude at the prescribed 3 mile point.

Figure 15-6

Formula for Computing Maneuvering Speed

$$V_a = V_s * \sqrt{g}$$

Where:

V_a = Maneuvering speed

V_s = Stall speed for configuration

g = Limit load factor

- Normal category = 3.8 g's
- Utility category = 4.4 g's

- Compute a maneuvering speed for maximum gross weight. This will be the HEAVY V_a.

- Compute a maneuvering speed for a gross weight representative of the aircraft with only the front seats occupied and no cargo or baggage. This will be the LIGHT V_a.

- When rough air dictates applying these V_a speeds, use a target speed 10 Kts less than the above speeds so that fluctuations in airspeed will not exceed the desired V_a speed.

Figure 15-7

Precision Visual Approach Traffic Pattern

This rate of descent will approximate the descent path of a 3 degree ILS glide slope.

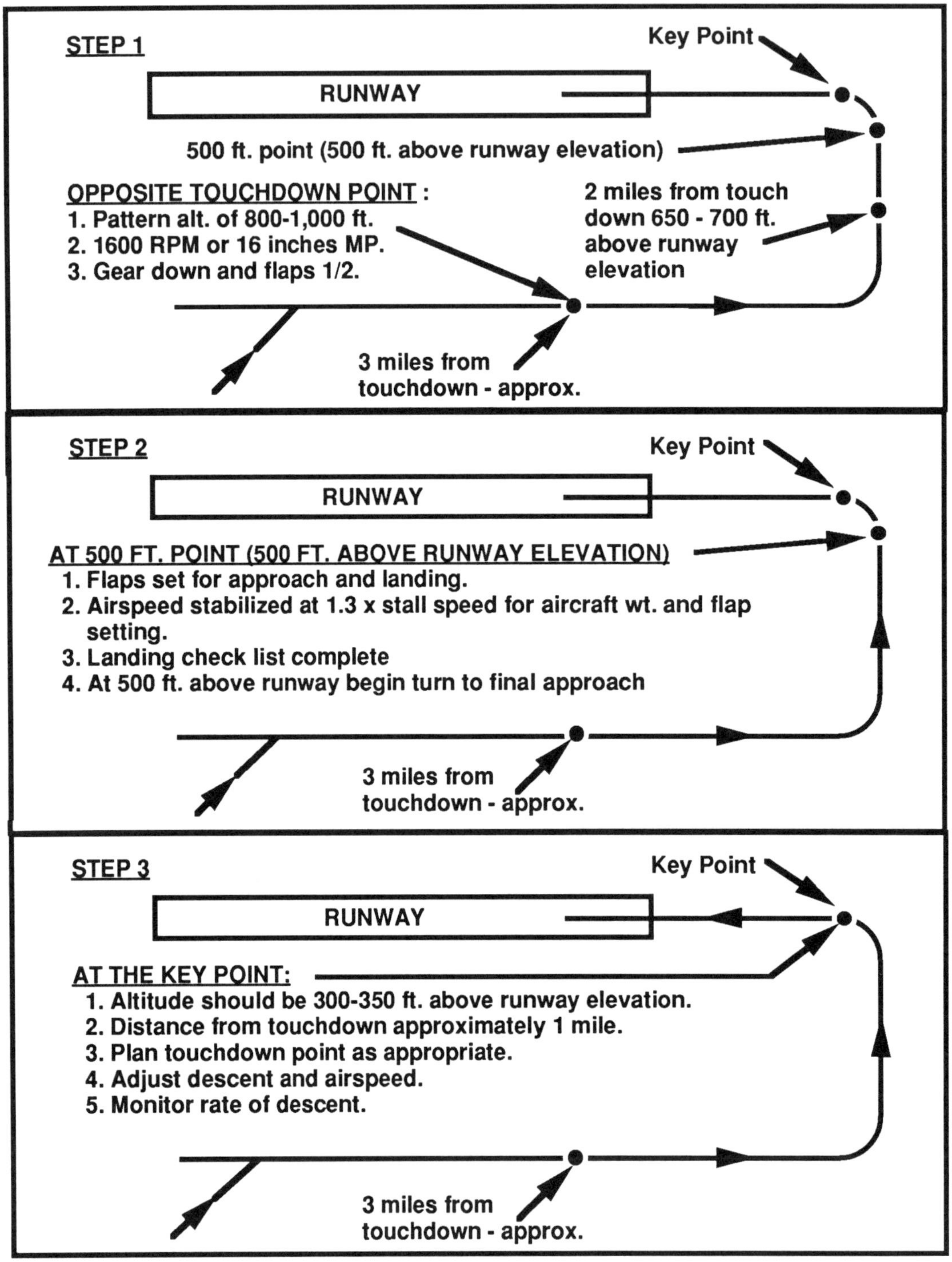

Figure 15-8

Precision Visual Approach
(Straight in Approach - VFR)

This rate of descent will approximate the descent path of a 3 degree ILS glide slope. Note that his approach is executed the same way as the traffic pattern approach in Figure 8-1.

STEP 1 - AT 1,000 FT. ABOVE RUNWAY ELEVATION

1. 1600 RPM or 16 inches MP.
2. Gear down and flaps 1/2 - approx.

STEP 2 - AT 500 FT. POINT (500 FT. ABOVE THE RUNWAY ELEVATION)

1. Flaps set for approach and landing.
2. Airspeed stabilized at 1.3 x stall speed for aircraft wt. and flap setting.
3. Plan touchdown point as appropriate.
4. Landing check list complete.
5. Monitor rate of descent.

STEP 3 - AT THE KEY POINT (300-350 FT. ABOVE THE RUNWAY ELEVATION)

1. Distance from touchdown approximately 1 mile.
2. Plan touchdown point as appropriate.
3. Monitor rate of descent.
4. Adjust airspeed and descent rate - as needed.

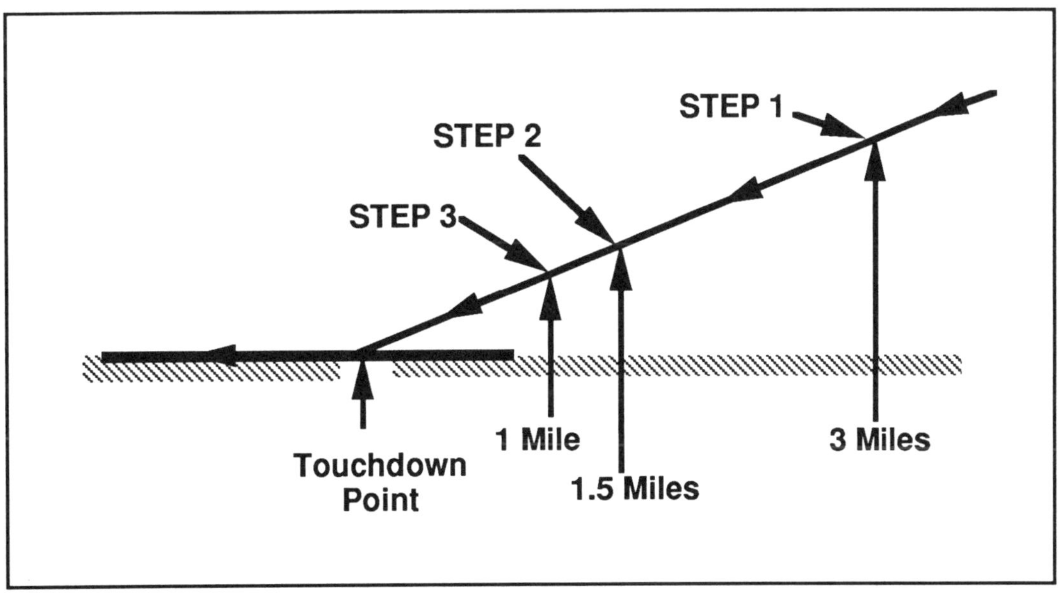

Figure 15-9

Do not blindly give command of your aircraft to the auto-pilot. I like former President Ronald Reagan's famous saying -- *"TRUST BUT VERIFY"* -- it is appropriate here! In other words know how the A/P works in all modes and monitor the operation as is reasonable to assure safe operation. Know the power source and all inputs to the system so that if things go wrong you are able to make an intelligent appraisal of what may be wrong and respond with appropriate corrective action.

When engaging the A/P or any of its features monitor the system response to assure that you see all the indications which are normal for the particular set up. In case the system commands an erroneous pitch attitude, or any abrupt and improper movement of the flight controls, the A/P should be overpowered and disengaged immediately. At this point let me mention that you should know the location of the circuit breakers for the A/P and the electric trim. These circuits and switches have been known to malfunction and *"runaway"* even though the disengage switch or control switch has been operated to the *"off"* position. In these unusual but frightening situations your only choice may be to deprive the system of electrical power by pulling the appropriate circuit breaker. In summary when the A/P system does not do what you want it to do, disengage it pronto and hand fly the airplane at the proper pitch attitude while you get things sorted out and corrective action taken.

Many pilots seem to be wrongly instructed regarding when to use PITOT HEAT. Pitot heat needs to be thought of as **preventative** equipment. Allowing the pitot system to ice up before turning it on is very unwise and in fact dangerous. WHAT ARE WE SAVING IT FOR?

Many fatal accidents have been caused by the pilot observing and responding to unbelievable airspeed indications which are caused by the pitot system being iced up. In most of such situations the pilot fixated on a weird airspeed indication while the airplane ran wildly toward destruction. These destructive results may be excessive IAS leading to airframe breakup or inadvertent stall.

Never engage the A/P unless **ALL** flight instrument indications are normal. For example -- suppose that we have just leveled off at cruise and are engaging the A/P for level cruise flight when we notice that the airspeed reads 60 Kts. or 245 Kts. Very bad. What to do? The drill here is to do two things -- set the **pitch attitude** to level flight attitude then assure that **cruise power** is being developed. If both these items are in place then you will be flying at or near cruise speed no matter what the airspeed indicator reads. Remember that **power** and **pitch** are what gets you home safely when all else fails. But, back to the problem at hand having the pitot heat turned on well in advance of need should be the regular practice.

My guideline on pitot heat is: Turn it on any time the O.A.T. is 40 degrees F or less and there is any precipitation or significant moisture present, or if there is any likelihood of encountering precipitation or cloud while in climb, cruise or any segment of the flight. It is worthwhile to note that the airline crews turn the pitot heat on at engine start and turn it off as they exit the landing runway. This professional view, of the need for pitot heat, is well worth our consideration, even though we are not an airline operation!

Use carburetor heat any time the need is indicated. We are taught to PULL CARB. HEAT ON EVERY APPROACH -- but every year many power failures or interruptions occur because the pilot failed to use carb. heat when it was needed during cruise or in the practice area. Carb. heat isn't just for approaches and landings -- it is for any time the need is anticipated or indicated. When the engine begins to slow down and or manifold pressure drops -- it is CARB. HEAT TIME!

If you fly an aircraft equipped with an alternate induction air system be sure you understand how the system operates in the manual and automatic mode and how to open the air induction air door manually should the need arise. These systems are not all the same so dig out the POH and learn all about the alternate air induction system on any airplane you fly. Do not bet on the door opening automatically because the door may become severely iced up making the manual control the only means for opening the door.

Study the POH for your aircraft at least once each 6 months. Go through this gold mine of information one page at a time. You will be surprised at the new information which you will discover each time you review the POH. There is good stuff between those covers of the POH. Dig it out.

Always have a generous fuel reserve and maintain a fuel burn out log while enroute.

Make a power check on every takeoff -- if things don't look normal -- DON'T GO!

Study again COCKPIT RESOURCE MANAGEMENT - Chapter 1. Concentrate on the management aspects of your flying.

Do not fail to get and understand a fully adequate weather briefing.

If there is any doubt about your cross country operating ability, be sure to get additional cross country training.

It was a black day in aviation when the FAA deleted spins from the Private Pilot license requirements. Find a good flight instructor and get some spin training. Around two hours will usually suffice. You need to be able to enter and recover from a two turn spin in each direction. Your confidence level will be enhanced and your orientation sharpened. It is fun. It takes some work but it is well worth the effort. This training may save your life!

Know the pitch attitude for climb at V_y speed and the pitch attitude for maintaining level flight when slightly below maneuvering speed (rough air speed). Find these numbers by flying your plane.

Remember -- all takeoffs and approaches -- day, night and IFR -- should use the same profile.

Make a habit of monitoring the rate of descent during **ALL** approaches. If you observe an indication either higher or lower than your usual *"ballpark"* indication (off by 200 FPM or more) you are in a very bad situation. Make a correction immediately. Be attentive to IVSI indications on approaches!

Utilize the precision visual approach plan until it becomes a natural procedure. It can save your life. It has saved my life and my airplane on many occasions over the years.

Lastly maintain a professional attitude. Be a responsible manager/pilot. Poor decisions are the root cause of most accidents even though that is not often the listed *"probable cause"*.

Manage like a pro. Fly like a pro. **Good Luck!**

NOTES:

NOTES: